D0025094

The Expansion of American Biology

Keith R. Benson
Jane Maienschein
Ronald Rainger
Editors

The Expansion of American Biology

Rutgers University Press
New Brunswick and London

Library of Congress Cataloging-in-Publication Data

The expansion of American biology / edited by Keith R. Benson, Jane Maienschein, and Ronald Rainger.
p. cm.
Includes bibliographical references and index.
ISBN 0-8135-1650-1—ISBN 0-8135-1651-X (pbk.)
1. Biology—United States—History. I. Benson, Keith Rodney. II. Maienschein, Jane. III. Rainger, Ronald, 1949–.
QH305.2.U6E97 1991
574′.0973—dc20 90-48070
CIP

Manufactured in the United States of America

To the memory of Brother Ed,
who exhorted us to "take care, stay well and keep smiling."
We miss him.

Contents

Contributors

Garland E. Allen is Professor of Biology at Washington University and the author of *Life Science in the Twentieth Century* (1975) and *Thomas Hunt Morgan: The Man and His Science* 1978).

John Beatty is Associate Professor of the History of Science and Technology in the Department of Ecology, Evolution, and Behavior, University of Minnesota.

Keith R. Benson is Associate Professor of Medical History and Ethics at the University of Washington. He edited *The American Development of Biology* (1988) with Ronald Rainger and Jane Maienschein and is currently completing a monograph on the history of marine biology laboratories in the United States.

Richard W. Burkhardt, Jr., is Professor of History at the University of Illinois and the author of *The Spirit of System* (1977). He is currently writing a book on the development of ethology as a scientific discipline, having already published several articles on the subject.

Adele E. Clarke is Assistant Professor of Sociology in the Department of Social and Behavioral Sciences and adjunct faculty in the Department of the History of Health Sciences at the University of California, San Francisco.

Hamilton Cravens is Professor of History at Iowa State University, Ames, where he teaches American cultural history and the history of science and technology. His most recent publication is a revised edition of *The Tri-*

umph of Evolution: The Heredity-Environment Controversy, 1900–1941 (1988).

Sharon E. Kingsland is Associate Professor in the History of Science at The Johns Hopkins University. She is the author of *Modeling Nature: Episodes in the History of Population Ecology* (1985).

Léo F. Laporte is Professor of Earth Sciences at the University of California, Santa Cruz. He also has a professional interest in the work of the American paleontologist George Gaylord Simpson.

Jane Maienschein is Professor of Philosophy and Zoology at Arizona State University. She is the author of *100 Years Exploring Life, 1888–1988,* a "biography" of the Marine Biological Laboratory, and coeditor with Ronald Rainger and Keith R. Benson of *The American Development of Biology* (1988).

Gregg Mitman is Assistant Professor in the Department of Ecology, Evolution, and Behavior, University of Minnesota.

Marilyn Bailey Ogilvie is an Associate Professor and Chair of the Division of Natural Science, Oklahoma Baptist University and an Adjunct Associate Professor of the History of Science, University of Oklahoma. She is the author of *Women in Science. Antiquity through the Nineteenth Century. A Biograhical Dictionary with Annotated Bibliography* (1986).

Diane B. Paul is Associate Professor in the Department of Political Science, University of Massachusetts at Boston.

Ronald Rainger is Associate Professor of History at Texas Tech University. He has edited, with Keith R. Benson and Jane Maienschein, *The American Development of Biology* (1988) and has published on the history of paleontology and evolutionary theory.

Preface

This volume is the second in a two-part series commissioned by the American Society of Zoologists to celebrate its centennial (1889–1989). As in the first volume, *The American Development of Biology*, we have tapped into the growing community of historians working in the history of twentieth-century American biology. As a result, all the chapters represent historical scholarship that has not been previously published and that, in fact, was prepared exclusively for this volume. We also used the same method of preparing contributions as before: the collaborators first wrote their essays and then circulated their submissions among one another; we all met together at Friday Harbor Laboratories to discuss our comments on the essays; and we subsequently revised the essays in several rounds of editing. We now offer *The Expansion of American Biology* to the reader, hoping that the whole of these collaborative efforts will prove to be greater than the sum of its parts.

We would like to thank many organizations and people for their support of this project. Once again, the American Society of Zoologists through its Centennial Committee provided generous support and Mary Adams-Wiley, the executive officer of the society, continually "greased the skids" when bureaucratic difficulties emerged. The Society's Division of the History and Philosophy of Biology also provided funds for the project, thus completing its pledge to the centennial of the society. Dean Dale Johnson of the Graduate School at the University of Washington provided additional funds from the Graduate School Research Fund. The Department of Medical History and Ethics contributed support for travel to and housing at Friday Harbor Laboratories. We also wish to thank Stephanie Fiskum and her daughter Haley for providing us with sumptuous meals and a feeling of being at "home away from home" during our conference at Friday Harbor. Without this financial and moral support, we could not have succeeded. Finally, Karen Reeds at Rutgers University Press has proved to be a never-ending source of encouragement and support. Her enthusiasm for the book has made all of our work much more tolerable. Our thanks to everyone.

The Editors

The Expansion of American Biology

Ronald Rainger

1 Introduction

In 1949 Warren Weaver, director of the Rockefeller Foundation's Natural Sciences Division, outlined his vision for biology.

> The century of biology upon which we are now well embarked is no matter of trivialities. It is a movement of really heroic dimensions, one of the great episodes in man's intellectual history. The scientists who are carrying the movement forward talk in terms of nucleoproteins, of ultra-centrifuges, of biochemical genetics, of electrophoresis, of the electron microscope, of molecular morphology, of radioactive isotopes. But do not be fooled into thinking that this is mere gadgetry. This is the dependable way to seek a solution of the cancer and polio problems, the problem of rheumatism and of the heart. This is the knowledge on which we must base our solution of the population and food problems. This is the understanding of life.[1]

Weaver's enthusiasm applied particularly to molecular biology, a term he had coined and a field he actively promoted. Yet his views also provide an important insight into some of the most significant changes occurring in twentieth-century American biology. In the years between 1920 and 1950 biology became "big science" and underwent profound changes in scope, methodology, and objectives. It became more closely integrated with physics, chemistry, and the social sciences and thus incorporated new methods, techniques, and instrumentation. And, as Weaver's comments illustrate, biology seemed to provide long-term solutions to a wide range of social problems. In brief, biology was expanding scientifically, methodologically, and in terms of its social and political applications.

The Size and Scope of Biology

Biology expanded considerably in strictly quantitative terms. By the end of World War I it was a well-established field of inquiry at a number of leading academic institutions. Studies by several scholars, including a number of essays in *The American Development of Biology,* have indicated that biology, though not yet a unified discipline in conceptual terms, had already gained departmental status at many colleges and universities, included a wide range of specialized disciplinary societies, and boasted many professional publications. Universities such as Chicago, Columbia, Harvard, and Johns Hopkins had research programs in genetics, biochemistry, and other fields that commanded worldwide attention. Independent institutions such as the Marine Biological Laboratory at Woods Hole, Massachusetts, were flourishing centers for research, education, and professional comraderie.[2]

Such expansion increased in the years after 1920. Whereas only 1.1 percent of all high school students were enrolled in a comprehensive biology course in 1910, biology had the highest enrollment of any field of secondary school science by 1950.[3] Teaching and research in biology also expanded tremendously within the context of higher education. Biology was originally concentrated at a few eastern institutions but later spread to colleges and universities throughout the United States. Of the more than 100 institutions of higher education that had conferred Ph.D. degrees in biology between 1950 and 1960, western and midwestern public universities were the leaders.[4] Important programs emerged at the University of California at Berkeley and Stanford, and the creation in 1928 of a department emphasizing molecular biology at the California Institute of Technology bore witness to the rise of western institutions as well as the development of a significant new field of inquiry. Other institutions gained prominence in different fields: a number of midwestern universities, including Chicago, Minnesota, and Nebraska, became centers for the study of ecology. Important programs in cytology and genetics developed at Indiana University. The University of Wisconsin became a leader in biochemistry.[5] The increasing emphasis on science in general no doubt spurred such growth and so did the reform in medical education. The vast overhaul of the nation's medical schools in the 1910s and 1920s, which systematized medical school curricula and mandated undergraduate education for all medical students, placed greater emphasis on biomedical studies and increased biology enrollments.[6]

Beyond sheer growth in size, biology also changed in more significant ways. Between 1920 and 1950 there were profound transformations in every field of biology. Developments in two new areas, genetics and biochemistry, have particularly captured the attention of historians. By 1920 American scientists were already at the cutting edge of research in those areas. Thomas Hunt Morgan and his associates at Columbia University had laid the foundations for the chromosomal theory of inheritance and transmission genetics, and subsequent research focused on understanding the structure of genes. According to standard interpretations, an expanding cadre of scientists, including physicists and chemists as well as biologists, developed the field of molecular biology by applying new techniques and instruments to the study of enzymes and proteins. That work, the traditional story suggests, culminated in James Watson and Francis Crick's discovery in 1953 that DNA is the carrier of hereditary information.[7] Other recent studies claim that a more complex and diverse set of factors characterize the origin and development of molecular biology. The influence of the work of the physicist Erwin Schroedinger has come under question. Other studies have challenged the assumed close working relationship among such leading figures as Schroedinger, Max Delbruck, and Watson.[8] The conceptual and institutional foundations of molecular biology have also received considerable attention. Robert E. Kohler has argued that molecular biology had its roots in a program promoted by Warren Weaver at the Rockefeller Foundation in the 1930s. In contrast, Pnina Abir-Am has maintained that an interdisciplinary group of European biologists, mathematicians, and physicists, although not entirely successful in obtaining funding, sought to ground biology in physics and chemistry. Lily E. Kay, in addition to providing important new insights into the work of Delbruck and W. M. Stanley, has defined the significant role that the California Institute of Technology played in the development of molecular biology.[9]

Much historical scholarship has also focused on biochemistry. A number of studies have concentrated on the intellectual history of biochemistry, tracing the historical development of scientific interpretations of biochemical structures and processes. In addition to the works of Robert Olby and Franklin H. Portugal, several studies have described the scientific research on proteins, enzymes, and various biochemical processes.[10] Others have explored the relationship of biochemistry to related fields and the impact of social, political, and philosophical factors on biochemical research. Robert E. Kohler has examined the relationship of biochemistry to medicine and chemistry and has defined the complex set of circumstances that affected its development as a distinct scientific discipline.[11] In addition to analyses of the work of Lawrence J. Henderson, recent biographical studies, notably Philip J. Pauly's work on Jacques Loeb and the biography of W. B. Cannon by Saul Benison, A. Clifford Berger, and Elin L. Wolfe, have enhanced our understanding of twentieth-century American biochemistry.[12] Those studies not only have described important researches, such as Loeb's invention of artificial parthenogenesis and interpretations of fertilization and proteins, but have also explained the work and careers of those individuals in relationship to the social, institutional, and intellectual context within which they operated. The biographies of Loeb and Cannon also help to define the relationship of biochemistry to physiology, particularly general physiology, as do several of the essays in Gerald L. Geison's recent edited volume, *Physiology in the American Context*.[13] In other associated fields Merriley Borell has examined the historical development of endocrinology out of a more generalized medical and physiological background. Her work, along with studies by Diana E. Long, has indicated the relationship of research in endocrinology to the birth control movement and the study of sex.[14] The discovery of vitamins has received some historical attention, particularly in relationship to studies of the history of nutrition.[15]

Genetics and biochemistry exploded into biology in the twentieth century, but important changes were occurring in the rest of biology, and on several levels. Virtually every field of biology experienced exponential growth in the sheer amount of data generated. Advanced research in biology required increasing specialization, and ecology, embryology, and systematics expanded into separate subdisciplines. Increasingly, laboratory analysis, more than fieldwork, became the sine qua non of biology. Along with laboratory analysis came more sophisticated instrumentation, methodology, and techniques. For example, not only did studies of the cell bring to the fore much new knowledge on cellular structure and function, but traditional studies of cell structure and function were accompanied by the use of biochemical, mathematical, and molecular analyses. In embryology, which had emphasized experimentation since the 1880s, new methods and interpretations emerged. Hans Spemann's concept of the organizer became the basis for a much improved understanding of the factors that determine development. Building on that work, Joseph Needham, Johannes Holtfreter, and others applied biochemical analyses and explanations to embryology.[16] Ecology and evolutionary biology, originally part of a natural history tradition, became increasingly mathematical and experimental. While Frederic Clements's use of the quadrat brought some quantification into ecology, the growth of population thinking and the application of sophisticated statistical and quantitative techniques in the work of Alfred J. Lotka and Vito Volterra placed ecology in a new and different context.[17] So too did the development of population genetics which, along with important studies by naturalists, played a major role in transforming the understanding of evolution and forging the evolutionary synthesis.[18]

Biology also became increasingly integrated with developments in other fields and employed new and different techniques. New questions, interpretations, and instruments were brought to bear on the study of molecular biology in the 1930s and 1940s,[19] but equally important, and less well known, was the integration occurring in other fields of biology. The study of animal behavior, long associated with psychology, not only drew upon interpretations from the social sciences but also adopted methods and techniques from biochemistry and endocrinology. Chemical analysis and techniques became more and more important for interpreting development as well as many other biological processes. Beginning in the 1940s cybernetics had a powerful impact on the understanding of inheritance and in later years was applied in such areas as the study of primate behavior. In those and many other ways the expanding subdisciplines within the biological sciences increasingly adopted and relied on interpretations, methods, and instrumentation employed in other sciences.[20]

The Many Uses of Biology

The objectives of biology and biologists also changed in the years 1920 to 1950. In brief, biology, as Weaver noted, received economic support in order to solve long-range social and political objectives. Long before 1920 scientists viewed their work in terms of its social significance. The nineteenth-century biologists William Keith Brooks, Charles Otis Whitman, and William Emerson Ritter, influenced by pragmatism and positivism, understood the importance of biology in terms of its social, political, and economic usefulness.[21] Biology also had specific applications in medicine, agriculture, husbandry, and even ecology. Since the mid-nineteenth century research in biochemistry and physiology had played an important role in medicine.[22] Seedsmen as well as scientists enthusiastically accepted Mendelian genetics, and practical as well as theoretical objectives promoted hybridization studies, especially the development of hybrid corn.[23] Explicit policy objectives directed the work of government biologists C. Hart Merriam, T. S. Palmer, and other members of the United States Biological Survey and the Bureau of Animal Industry. Working in close cooperation with ranching and farming interests, they sought to control animal populations, particularly populations of predatory animals, through the development of poisons and other forms of eradication.[24] Eugenics, an attempt to control human population and behavior, achieved widespread popularity in the 1910s and early 1920s. A number of biologists promoted birth control measures and others contributed to the passage of an immigration restriction bill.[25]

Still there were some subtle, though significant, changes that occurred after 1920. Following World War I, philanthropic foundations provided much greater financial support for biology. In addition, those organizations underwrote research that was expected to have a social return only in the long run, in contrast to earlier objectives that anticipated that directly applied research would provide immediate solutions to social problems.

The increased support of biological research was in part a manifestation of the changing role and status of American science. By the mid-twentieth century the United States had become the world's leading power. The victor in two world wars, it had unrivaled political and economic influence. By

1950 the United States was producing one-half of the world's manufactured goods and was actively promoting a foreign policy designed to extend political influence and expand economic opportunities. Many people, particularly in the years before 1945, equated science and engineering with progress. To a mass consumer society that valued and purchased automobiles, radios, vacuum cleaners, and dozens of other products in staggering numbers, science and technology seemed to provide the promise of ever-increasing prosperity and growth. Utopian writers envisioned a world in which the wonders of science and technology would transform civilization. In the 1930s a small group of technocrats called for engineers to take control of government and politics.[26] Certainly not everyone had such an abiding faith in contemporary developments. Writers ranging from the theologian Reinhold Neibuhr to the scientists Robert Millikan and Henry Fairfield Osborn decried the rise of "modern civilization" and sought to preserve older values and institutions. Yet Millikan and Osborn also employed science as a means for solving what they identified as contemporary social, economic, and political problems.[27]

The expanding faith in science and technology played a part in the development of greater funding and patronage for science. Prior to 1920 the sciences in the United States received little support from the government or other institutions. Although the Carnegie Institution of Washington and various Rockefeller agencies had begun to provide some assistance for science by 1915, it was only during and after World War I that the enterprise received significant attention and support. The development of the National Research Council in 1916 provided government support for wartime scientific research. Biology, except perhaps through agriculture, had few immediate applications, but chemists and physicists developed sonar, flash- and sound-ranging devices, and the means for combating and waging chemical warfare that aided in the wartime effort.[28] After the war several philanthropic organizations, including the Rockefeller Foundation, the Carnegie Corporation, the Russell Sage Foundation, and the Guggenheim Foundation, began to provide considerable support for biology and other sciences.

The support of science by private philanthropies was a consequence not solely of its increased visibility but also of a belief that science could solve problems and promote advancement. "The growth of philanthropic support for medicine, natural science, and then social science represented a certain pattern of belief about the value of 'scientific knowledge' in the good life and the good society."[29] Drawing on the belief that the application of scientific method and the use of scientific knowledge could improve civilization, private organizations, particularly a number of Rockefeller organizations, supported the sciences in a wide variety of ways. Many of those programs focused on medicine and public health. The Rockefeller Foundation Sanitary Commission worked to eradicate hookworm. The International Education Board, the General Education Board, and the Laura Spelman Rockefeller Memorial promoted medical education as well as numerous public health and social welfare projects. The Rockefeller Foundation also promoted new objectives for agriculture, "breeding miracle grains for the green revolution."[30] Through the Rockefeller Foundation and other institutions biology in the interwar years was undergoing a transition "to something which involves the mass application of an industry."[31]

Beyond applying science to solve practical problems, individual scientists and patrons also conceived of their work as a means for solving other issues, including social and political problems. Gregg Mitman has argued that the tragic consequences of World War I led a number of biologists to address social and political problems directly through their scientific work. Many biologists, Mitman maintains, revamped their interpretations of evolution, particularly the commitment to competition and natural selection, in response to the events of World War I. Mitman has also argued that social assumptions and ideals played a central role in the ecological studies of Warder Clyde Allee and Alfred E. Emerson. William Albury and Stephen J. Cross have claimed that the models for biochemical regulation developed by W. B. Cannon and Lawrence J. Henderson in the 1930s were organic analogies that had specific political meaning for solving the problems of the Depression.[32] Technocratic models, according to Peter J. Taylor, influenced the systems ecology of H. T. Odum, and Donna Haraway has maintained that the analysis of social roles and behavior among primates was closely intertwined with concerns for assessing, evaluating, and controlling human organization and behavior.[33]

Organizations such as the Rockefeller Foundation embodied such commitments. Recent studies have questioned the argument that foundations constituted a means to propagate capitalism and self-interest.[34] However, the Rockefeller Foundation and other agencies sponsored programs to promote human social as well as biological welfare and did so within a framework that accepted a commitment to the existing social, economic, and political order. In the 1930s and 1940s scientists and their patrons moved away from racist assumptions and the search for immediate solutions that characterized the early eugenics movement. Instead philanthropic organizations, as well as the government, sponsored programs on human biology, population control, and racial relations that sought long-term solutions to human social and biological problems.[35] Those programs provided for expansion of scientific research in several fields of biology, and a changing perception of the social and political usefulness of that research.

Overview of the Essays

These themes, the emphasis on biology's increasing size and scope, specialization, integration, and application for the purposes of practical benefit and social control, are brought forth in new and important ways in the essays in this volume. The volume does not pretend to cover all aspects of American biology in the period 1920–1950. There are, for example, no essays that deal explicitly with molecular biology and biochemistry. There is already an extensive literature on the history of those subjects. Moreover, molecular biology and biochemistry primarily came into biology from without, through the efforts of physicists, chemists, and others, and quickly developed their own distinct disciplines. The essays in this volume, which are previously unpublished and original, mine new ground by analyzing the changing developments and expansion within the various subdisciplines of biology. Through such an analysis the essays in this book not only establish important ties to the preceding volume, *The American Development of Biology,* but also provide the most effective means for understanding and evaluating conceptual, methodological, social, and interdisciplinary developments that occurred in biology during the first half of the twentieth century.

A number of the essays address the theme of expansion by examining particular works or individuals. The increasing specialization within biology forms the subject of Jane Maienschein's essay. Focusing on a multiauthored text, E. V. Cowdry's *General Cytology* (1924), Maienschein argues that although Edmund Beecher Wilson's seminal work, *The Cell in Development and Heredity,* had long dominated the field, by the 1920s cytology encompassed such an extensive body of data and such a wide variety of methods and techniques that no one individual or no one approach could adequately cover the subject. Maienschein examines the changing developments in that field through a comparative analysis of the Cowdry volume and the third edition of Wilson's book (1924). The latter, a monumental work, concentrated on cellular morphology; in contrast, the Cowdry volume comprised specialized essays on cellular chemistry, cellular physiology, cellular behavior, fertilization, and a host of other subjects. The essays in that volume reflected an expanding range of questions and a diversity of new techniques, including tissue culture, microdissection and injection, and cellular reactions to electric stimuli, that characterized the work of the country's leading students of the subject. By the 1920s, Maienschein demonstrates, the study of cytology required a variety of perspectives, techniques, and emphases. Her essay also suggests other developments pertaining to the expansion occurring within biology. The publication of such a multiauthored text indicated a certain self-awareness on the part of biologists, the recognition that biology was a rapidly expanding field and as such required new means of conveying information. Implicitly, too, that volume illustrated the expanding market in biology and biology education which required a comprehensive, multiauthored text.

Whereas Maienschein's contribution examines two texts as the means for delineating the wide range of changes affecting early twentieth-century biology, Marilyn Ogilvie's study provides a perspective on still other dimensions of expansion: the role of women in biology and the exportation of American biology beyond its continental borders. Examining two prominent women biologists, Nettie Maria Stevens and Alice Middleton Boring, Ogilvie argues that changes in American universities, while seemingly opening new opportunities, also created challenges and constraints that affected the careers and the work of women scientists. Stevens, who completed her Ph.D. with Thomas Hunt Morgan, obtained a teaching position at Bryn Mawr but experienced difficulties in achieving professional advancement or financial support for her research. Boring, who studied with Stevens, pursued research in cytology and genetics at the University of Würzburg, the Naples Zoological Station, and the University of Maine. In the late 1910s, however, she left the position at Maine to take up teaching and a different field of biological research in China. Financial support from the Rockefeller Foundation provided new teaching and research opportunities in that country, particularly at Peking Union Medical College. But, as Ogilvie argues, gender restrictions, logistical problems, and political unrest in China placed constraints on experimental research that had important consequences for Boring's work. Boring devoted herself to teaching in China but her research interests shifted from cytology and genetics to reptile taxonomy. That work, by providing important new faunal data, transcended the mere exportation of experimental emphases and techniques. Taxonomic work also fit in more readily with the political, financial, and time limitations that Boring experienced in China. Ogilvie's essay describes Boring's work as a teacher and a taxonomist, but more significant, it demonstrates the constraints on women biologists in the early- and mid-twentieth century. Biology was expanding to include larger numbers of women, but in occupational terms they remained underemployed. In Boring's case a noted woman biologist found opportunity and satisfaction only through a radical change of career goals and interests.

Léo Laporte's essay focuses on one individual to illustrate expansion. His study emphasizes the role that George Gaylord Simpson played in bringing paleontology into line with contemporary work in genetics and evolutionary biology. Simpson, Laporte argues, began to employ population concepts and statistical methods in paleontological studies in the late 1930s. His book, *Tempo and Mode in Evolution* (1944), was one of the major texts of the evolutionary synthesis. In that work Simpson applied the methods and interpretations of population genetics to the fossil record, and, as Laporte emphasizes, that book played a seminal role in leading paleontologists to change from a typological to a population concept of species. Laporte also breaks new ground by indicating how Simpson's work introduced paleontologists to ecological concepts. In an important respect Simpson was working to change and to expand the methods and interpretations that dominated early twentieth-century paleontology. He was incorporating concepts from population genetics and ecology as well as statistical methods and techniques into paleontology. According to Laporte, Simpson's work also reflects expansion and the changing nature of biology in another respect: Simpson's awareness of the important role he was playing. Cognizant of the status of paleontology in the early 1930s, Simpson was aware of the need to introduce major theoretical changes into the discipline to bring paleontology into the mainstream of modern evolutionary biology. He self-consciously took up that task. Although he played a central role in integrating paleontology into biology, Simpson also continued to emphasize and to work for disciplinary independence for paleontology. As such, his work reflects an increasing understanding of the role of paleontology and of its relationship to biology, a greater sense of definition of what constitutes paleontology and biology.

Expansion in biology also took place through the incorporation of work in other fields into biology. Adele Clarke's essay examines that subject by describing the development of a whole new field of research: the reproductive sciences. Historians have defined the separation of genetics and embryology from an older core of research problems.[36] However, Clarke argues that reproductive science also emerged as a distinct, specialized field from the same origins. By the 1910s F. H. A. Marshall and Frank Rattray Lillie defined reproductive physiology as a set of problems that could be addressed on its own. Clarke demonstrates that while Marshall and Lillie approached the study of reproduction from the perspective of embryology, the reproductive sciences also drew on developments occurring in the new field of endocrinology, which was itself becoming distinct from physiology. Through an analysis of Lillie's research on the freemartin and Charles Stockard and George Papanicolaou's examination of the estrus cycle of the guinea pig, Clarke describes some of the most significant, early work in reproductive science. Both studies were driven by embryological problems, but both also indicated the importance of hormonal processes in sexual function. Clarke argues that that work, particularly the development of the "Pap Smear," had important results: it stimulated research on reproductive problems, promoted further collaboration between biologists and biochemists, and influenced work in controlling the reproductive process. By the 1930s the reproductive sciences had emerged as a legitimate and flourishing field of biological research.

Other essays examine the close relationship between biology and the social sciences. Hamilton Cravens's contribution, "Behaviorism Revisited," addresses that topic by describing a transformation in the understanding of behavior and learning. According to Cravens, psychologists and biologists in the 1910s and 1920s began to replace an understanding of organism and behavior based on the study of discrete parts and average types, with a different interpretation of natural and social reality that emphasized the whole as greater than the sum of its parts. Cravens focuses on the early work of John B. Watson who in the 1910s developed an interpretation of behavior characterized by an understanding of organisms and organic processes as dynamic, complex interactions among interdependent parts. Watson had an abbreviated academic career and left an ambiguous legacy, though Cravens describes the important impact that Watson's interpretation had on many biologists and psychologists of the next generation. Leonard Carmichael and George Ellett Coghill, both of whom did research on amphibians, developed theories of maturation that embodied Watson's ideas. The postulates for maturation theory that they worked out for lower vertebrates became a model for animal behaviorists and child psychologists in the 1930s. Experiments carried out by Arnold L. Gesell, Nancy Bayley, and a host of others demonstrated that skill, performance, intelligence, indeed almost any trait, developed through the interaction of heredity and environment, specifically the maturation of the central nervous system and the integration of its distinct parts into a larger whole. In addition to offering a new interpretation of Watson's influence, Cravens's analysis of the work done on learning and behavior indicates the pervasiveness of interdisciplinary research in mid-twentieth-century biology: an extensive series of intellectual and institutional networks developed between biologists and psychologists and were brought to bear on problems of child and animal behavior.

The work of animal behaviorists is also the subject of the essay by Gregg Mitman and Richard W. Burkhardt, Jr. Through an analysis of the researches of Warder Clyde Allee, Gladwyn Kingsley Noble, and Margaret Morse Nice, their essay identifies the diverse traditions that characterized the study of animal behavior. In contrast to Nice who pursued field studies, Allee and Noble concentrated on laboratory research and experimentation to understand behavior. Noble incorporated methods and interpretations from biochemistry and neurology to investigate animal behavior. Concentrating on problems of reproductive physiology, he employed neural surgery and hormonal analysis to examine animal social and sexual behavior, techniques that reflected the growing importance of endocrinology and reproductive physiology as discussed by Clarke. Mitman and Burkhardt indicate that Allee was also interested in the origins of social behavior and employed hormonal and vitamin experiments to explore questions of social rank, dominance, and territoriality. More important, Allee's work focused on community animal ecology and drew heavily from a tradition of sociological research at the University of Chicago. Interested in understanding the dynamics of animal aggregates, Allee framed his research and interpretations within the context of a theory of sociality. Mitman and Burkhardt maintain that Allee's interpretation of animal behavior, which emphasized cooperation among animal aggregates, was intertwined with and had ramifications for his understanding of human social behavior.

The social and political dimensions of behavioral studies are more fully developed in their analysis of Nice. Technically an amateur who operated outside the academic context, Nice nonetheless played a prominent part in the development of the subject. Through her reviews of current ornithological literature she made the latest work on birds available to an interested public. Through her extensive personal correspondence Nice brought the work of important European ethologists to the attention of American scientists. She also prodded and challenged Noble and others to respond to the new interpretations being advanced by Konrad Lorenz. Her life histories of bird families were important in their own right, though, as Mitman and Burkhardt point out, her career illustrates the professional constraints placed on Nice as a result of gender. Her work, particularly her increasing skepticism about the importance of male dominance theory, indicates the links between gender and behavioral interpretations.

Sharon Kingsland's essay examines the efforts to develop a biology of human behavior by analyzing the work of Charles Manning Child and Charles Judson Herrick. Colleagues at the University of Chicago, Child and Herrick helped establish a tradition in Chicago biology that emphasized a dynamic conception of the organism and rejected reductionism. Both Child and Herrick, Kingsland argues, combined experimentation with a natural history orientation to the study of the organism, a commitment not entirely dissimilar to the behavioral interpretations that Cravens emphasizes. Seeking to combine morphology with other studies, notably physiology and neurology, they developed an interpretation and approach that emphasized examining the organism in relationship to its environment. Child's work, which focused on dominance relationships and influenced his colleagues in ecology at Chicago, rejected the specialization and reductionism associated with classical genetics. Herrick, influenced by Child and by his brother's interest in psychobiology, developed an interpretation of neurological behavior very similar to Child's interpretation of animal physiology. For Herrick, as for Child, dominance relationships among organic parts and the relationship of individual to environment were the basis for understanding behavior. Kingsland maintains that Child and Herrick developed a dynamic, evolutionary perspective that embodied a commitment to holism and led them to embrace the ideas of emergent evolution advanced in the 1920s. Her essay also ties their scientific views to the liberal political ideology and democratic philosophy of John Dewey's pragmatism. Thus Kingsland, as well as Mitman and Burkhardt, demonstrate the close relationship between scientific and political discourse that existed in the work of many American biologists in the years after 1920.

Yet, as Kingsland's essay emphasizes, Herrick's work illustrates another important feature of mid-twentieth-century American biology: the concern for social control. Kingsland and Mitman and Burkhardt, in line with other recent work in the history of science, indicate that the social, political, and economic troubles of the 1920s and 1930s led many biologists to construct interpretations that had overt social and political meaning. The devastation of World War I, Mitman and Burkhardt argue, influenced Allee, a Quaker, to define the relationships within animal and human communities in terms of cooperation. Herrick explicitly sought to understand human behavior in order to control individual activity and social movements. In place of a strictly laissez-faire system, he maintained that cooperation and planning, particularly by social control through education, would give rise to a meritocracy capable of handling society's problems. Traditionally, historians have identified an interest in a biology of social control as a consequence of reductionism in molecular biology and biochemistry, or as a manifestation of the public policies of government biologists. But, as Kingsland demonstrates, an interest in biological research as a means for regulating human behavior was also embedded in the holistic interpretations of Herrick and Child.

The application of biological research to achieve political or social objectives is further delineated in several other essays in the volume. One of the foremost efforts along that line was eugenics, a movement that historians have traditionally argued tapered off in the mid or late 1920s. However, Garland Allen, in his essay on Raymond Pearl, argues that eugenics was not so much abandoned as transformed into a concern for population control. Pearl, an early supporter of the hereditarian eugenics of the 1910s, became increasingly critical of that movement but did not abandon his elitist views or his concern for regulating human behavior. Instead, in the 1930s he took up the cause of population control, of studying and regulating reproduction, particularly the problems pertaining to the proliferation of what he defined as the defective segment of human society, overpopulation, and human migration. Pearl, according to Allen, was not only the leading student of population control but also a popularizer and promoter of the cause. From 1925 to 1930 he obtained support from the Rockefeller Foundation and later from the Milbank Memorial Fund for a population control research center at Johns Hopkins University. Allen indicates that biometrics, not Mendelian genetics, was the basis for Pearl's interest in eugenics and population control; in that respect he was somewhat different from other eugenicists. Yet Allen's essay suggests that, in addition to Pearl, the promotion of population control studies by several individuals and organizations in the late 1930s and 1940s indicates a continuing concern with the regulation of human biology and behavior.

Diane Paul's essay further illustrates the social objectives that influenced work in genetics. She demonstrates that the Rockefeller Foundation, well known for its role in promoting molecular biology, was a major sponsor of behavior genetic research as well. Under the leadership of Alan

Gregg, director of the Rockefeller Foundation's Medical Sciences Division, that institution funded a number of projects on the genetics of behavior, including a major long-term study of emotional and intellectual variation in dogs. Gregg hoped the study would prove that the source of this variation was largely genetic rather than environmental. In his view, such a demonstration would have obvious implications for the human "nature-nurture" debate and thus for medical and educational practice. In the end, however, the results of the study contradicted his assumptions and were effectively buried by the foundation. Paul's essay, along with Allen's and Ogilvie's, not only indicates the role that the Rockefeller Foundation and other private agencies played in supporting biology in the mid-twentieth century, but, more important, shows that the political and social objectives of patrons had a powerful impact on biological research.

Political objectives also played a part in genetics research in the aftermath of World War II, as John Beatty's essay illustrates. Following the bombings of Hiroshima and Nagasaki, the National Academy of Sciences, with funding from the newly created Atomic Energy Commission, established the Atomic Bomb Casualty Commission. Among its many endeavors the commission developed a genetics project designed to determine the genetic effects of exposure to atomic bomb radiation among Japanese survivors. The genetics project, organized and directed by biologist James Neel, explicitly included the possibility that no genetic effects, in other words the null hypothesis, could not be ruled out by the study. No genetic effects was the conclusion reached by the commission, and while negative results generally have little significance in scientific research, and generally are not reported or published in science journals, Beatty argues that in this case such results had considerable relevance and importance. In the years immediately following World War II, when the United States needed Japanese support and faced cold war tensions on a global scale, the conclusion that the atomic bomb explosions over Japan had no significant genetic effect had demonstrable political value and significance. Beatty also argues that that conclusion had important consequences in domestic affairs. In the early 1950s, when the United States was developing a vast nuclear weapons arsenal, testing atomic bombs above ground, and inaugurating a nuclear energy program, the commission's findings were important. Beatty's essay does not suggest that the commission's genetics project was explicitly designed and carried out to reach politically expedient conclusions. Nevertheless, by indicating the unusual importance accorded negative results, Beatty points out the important bearing that political objectives had on the genetics research conducted by the Atomic Bomb Casualty Commission.

The commission's work also had significant scientific consequences. The study by Neel and William Schull was criticized by other geneticists, notably Hermann J. Muller and Alfred H. Sturtevant, who maintained that humans were not appropriate subjects for genetic research. Neel and Schull had to refute those arguments and demonstrate that humans were suitable material for genetic research and would need to be included in any future radiation studies. Through that analysis Beatty's essay, as well as Paul's, identifies some of the important early efforts and some of the difficulties in establishing the legitimacy of human genetics.

Conclusion

The essays in this volume demonstrate the changing nature of biology in the period 1920–1950 and do so in a manner that offers important new perspectives on the history of American biology. From an underfunded, largely academic endeavor in the years around 1920, biology by 1950 had become a vast, well-endowed enterprise of considerable scientific as well as social and political significance. As several of these essays illustrate, biologists became increasingly aware of the need to institute conceptual and methodological changes in the science and realized that the application of their interpretations had the potential for providing long-range solutions to contemporary problems. Philanthropic organizations and the government adopted similar views, and in the years after World War I they supported and promoted work in the biological sciences on a scale that was unprecedented. Building on current historiographical emphases, these essays examine not only important dimensions of biological research but also the relationship of that research to social, political, and philosophical issues. Through the analysis of the work of prominent individuals and institutions they describe the exchange and interaction of ideas and methods across disciplinary boundaries, the emergence of new fields of research, the bearing of social, political, and economic concerns on science, and the changing objectives of biology in the mid-twentieth century. In ways not previously developed, the essays in this volume illustrate the expanding size, scope, and social role of biology in mid-twentieth-century America.

Acknowledgments

I thank the participants in the Friday Harbor Conference for their instructive comments on an earlier draft of this essay.

Notes

1. Warren Weaver to H. M. H. Carlson, 17 June 1949, quoted in Raymond B. Fosdick, *The Story of the Rockefeller Foundation* (New York: Harper, 1952), p. 166.
2. See the following essays in Ronald Rainger, Keith R. Benson, and Jane

Maienschein, eds., *The American Development of Biology* (Philadelphia: University of Pennsylvania Press, 1988): Keith R. Benson, "From Museum Research to Laboratory Research: The Transformation of Natural History into Academic Biology," pp. 49–83; Toby A. Appel, "Organizing Biology: The American Society of Naturalists and Its 'Affiliated Societies,' 1883–1923," pp. 87–120; Philip J. Pauly, "Summer Resort and Scientific Discipline: Woods Hole and the Structure of American Biology, 1882–1925," pp. 121–150; and Jane Maienschein, "Whitman at Chicago: Establishing a Chicago Style of Biology?" pp. 151–182.

3. William V. Mayer, "Biology Education in the United States During the Twentieth Century," *Qrtly. Rev. Biol.*, 1986, *61*: 483–484. Joseph D. Novak, *The Improvement of Biology Teaching* (Indianapolis: Bobbs Merrill, 1970), figure 1-2, pp. 6–7, shows that by 1950 over 1.6 million students took classes in biology during the four years of public high school. By comparison, 250,000 students took courses in physics and 700,000 in chemistry. Table 1–1, pp. 8–9 of the same work, shows that as of 1949 18.4 percent of public high school students took classes in biology, as distinct from classes in physiology, 1 percent, zoology, .1 percent, and botany, .1 percent. Again biology far outstripped physics and chemistry which attracted 5.4 percent and 7.6 percent of high school students, respectively. Among other leading surveys that illustrate the expanding role of biology in secondary school education, see Oscar Riddle et al., *The Teaching of Biology in the Secondary Schools of the United States* (New York: Union of American Biological Societies, 1942).

4. Clarence B. Lindquist and Edwin L. Miller, "Degrees Conferred in the Biological Sciences," *AIBS Bulletin*, 1962, *12*: 28–32, on p. 31, table 4.

5. On the University of California at Berkeley, see Richard M. Eakin, "History of Zoology at the University of California at Berkeley," *Bios*, 1956, *27*: 62–90. On Stanford, J. Percy Baumberger, "A History of Biology at Stanford University," *Bios*, 1954, *25*: 123–147. Developments in biology at the California Institute of Technology are discussed in Garland E. Allen, *Thomas Hunt Morgan: The Man and His Science* (Princeton: Princeton University Press, 1978), pp. 344–400; Robert Olby, "The Origins of Molecular Biology at Cambridge and Caltech," in *Proceedings of a Conference on the History of Bioenergetics* (Boston: American Academy of Arts and Sciences, 1975), pp. 60–96; and most fully in Lily E. Kay, "Cooperative Individualism and the Growth of Molecular Biology at the California Institute of Technology" (Ph.D. dissertation, Johns Hopkins University, 1987). The development of ecology at the University of Nebraska is the subject of Ronald E. Tobey, *Saving the Prairies: The Life Cycle of the Founding School of American Plant Ecology, 1895–1955* (Berkeley: University of California Press, 1981). On biology at Indiana University, see the discussion of the work of H. J. Muller in Elof Axel Carlson, *Genes, Radiation and Society: The Life and Work of H. J. Muller* (Ithaca: Cornell University Press, 1982), pp. 289–303; and the discussion of the work of T. M. Sonneborn in Jan Sapp, *Beyond the Gene: Cytoplasmic Inheritance and the Struggle for Authority in Genetics* (New York: Oxford University Press, 1987), pp. 87–122. The significance of biochemistry at the University of Wisconsin is discussed in John T. Edsall and David Bearman, "Historical Records of Scientific Archives: Survey of the Sources for the History of Biochemistry and Molecular Biology," *Proceedings of the American Philosophical Society*, 1979, *123*: 279–292, on pp. 286–288.

6. Robert E. Kohler, *From Medical Chemistry to Biochemistry: The Making of a Biomedical Discipline* (Cambridge: Cambridge University Press, 1982), pp. 121–157; and Kenneth M. Ludmerer, *Learning to Heal: The Development of American Medical Education* (New York: Basic Books, 1985).

7. Robert Olby, *The Path to the Double Helix* (London: Macmillan, 1974); Horace Freeland Judson, *The Eighth Day of Creation: Makers of the Revolution in Biology* (New York: Simon and Schuster, 1979); Franklin H. Portugal and Jack S. Cohen, *A Century of DNA: A History of the Discovery of the Structure and Function of the Genetic Substance* (Cambridge, Mass.: MIT Press, 1977).

8. E. J. Yoxen, "Where Does Erwin Schroedinger's *What is Life* belong in the History of Molecular Biology?" *Hist. Sci.*, 1979, *17*: 17–51. Pnina Abir-Am, "Themes, Genres, and Orders of Legitimation in the Consolidation of New Scientific Disciplines: Deconstructing the Historiography of Molecular Biology," *Hist. Sci.*, 1985, *23*: 73–117, esp. pp. 83–86.

9. Robert E. Kohler, "The Management of Science: The Experience of Warren Weaver and the Rockefeller Foundation Programme in Molecular Biology," *Minerva*, 1976, *14*: 279–306; idem, "Warren Weaver and the Rockefeller Foundation Program in Molecular Biology: A Case Study in Scientific Management," in Nathan Reingold, ed., *The Sciences in the American Context: New Perspectives* (Washington, D.C.: Smithsonian Institution, 1979), pp. 249–293. Pnina Abir-Am, "The Discourse of Physical Power and Biological Knowledge in the 1930s: A Reappraisal of the Rockefeller Foundation's 'Policy' in Molecular Biology," *Social Studies of Science*, 1982, *12*: 341–382; idem, "The Biotheoretical Gathering, Transdisciplinary Authority and the Incipient Legitimation of Molecular Biology in the 1930s: New Perspective on the Historical Sociology of Science," *Hist. Sci.*, 1987, *25*: 1–70; idem, "The Assessment of Interdisciplinary Research in the 1930s: The Rockefeller Foundation and Physico-chemical Morphology," *Minerva*, 1988, *26*: 153–176. Lily E. Kay, "Conceptual Models and Analytical Tools: The Biology of Physicist Max Delbruck," *J. Hist. Biol.*, 1985, *18*: 207–246; idem, "W. M. Stanley's Crystallization of the Tobacco Mosaic Virus, 1930–1940," *Isis*, 1986, *77*: 450–473.

10. See Garland E. Allen, *Life Science in the Twentieth Century* (Cambridge: Cambridge University Press, 1978), pp. 147–185. Also see William Bechtel, "Biochemistry: A Cross-disciplinary Endeavour that Discovered a Distinctive Domain," in William Bechtel, ed., *Integrating Scientific Disciplines* (Dordrecht: Reidel, 1986), pp. 77–100; Marcel Florkin, *A History of Biochemistry* (Amsterdam: Elsevier, 1972); Joseph F. Fruton, *Molecules and Life: Historical Essays on the Interplay of Chemistry and Biology* (New York: Wiley-Interscience, 1973); John T. Edsall, "Proteins as Macromolecules: An Essay on the Development of the Macromolecular Concept and Some of Its Vicissitudes," *Archives of Biochemistry and Biophysics*, 1962, *1* (suppl.): 12–20; idem, "The Evolution of Knowledge of Functional Adaptation in a Biochemical System. Part 1. The Adaptation of Chemical Structure to Function in Hemoglobin," *J. Hist. Biol.*, 1972, *5*: 205–257; idem, "Carbon Dioxide Transfer in Blood: Equilibrium between Red Blood Cells and Plasma: The Work of D. D. Van Slyke and L. J. Henderson, 1920–1928," *History and Philosophy of the Life Sciences*, 1985, *7*: 105–120; Frederic L. Holmes, "Joseph Barcroft and the Fixity of the Internal Environment," *J. Hist. Biol.*, 1969, *2*: 89–122; Robert E. Kohler, "The

Enzyme Theory and the Origin of Biochemistry," *Isis,* 1973, *64*: 181–196; P. R. Srinivasan, Joseph Fruton, and John T. Edsall, eds., "The Origins of Modern Biochemistry: A Retrospect on Proteins," *Annals of the New York Academy of Sciences,* 1979, *325*: 1–360; Mikulas Teich, "The History of Biochemistry," *Hist. Sci.,* 1975, *17*: 193–245.

11. Kohler, *From Medical Chemistry to Biochemistry.*

12. On L. J. Henderson, see John Parascandola, "Organismic and Holistic Concepts in the Thought of L. J. Henderson," *J. Hist. Biol.,* 1971, *4*: 63–113; and Stephen J. Cross and William R. Albury, "Walter B. Cannon, L. J. Henderson, and the Organic Analogy," *Osiris,* 1987, *2nd ser. 3*: 165–192. For biographic studies, see Philip J. Pauly, *Controlling Life: Jacques Loeb and the Engineering Ideal in Biology* (New York: Oxford University Press, 1987); Saul Benison, A. Clifford Berger, and Elin L. Wolfe, *Walter B. Cannon: The Life and Times of a Young Scientist* (Cambridge, Mass.: Belknap Press of Harvard University Press, 1987).

13. See the following essays in Gerald L. Geison, ed., *Physiology in the American Context, 1850–1940* (Bethesda, Md.: American Physiological Society, 1987): Toby A. Appel, "Biological and Medical Societies and the Founding of the American Physiological Society," pp. 155–176; Jane Maienschein, "Physiology, Biology, and the Advent of Physiological Morphology," pp. 177–193; and Philip J. Pauly, "General Physiology and the Discipline of Physiology, 1890–1935," pp. 195–207.

14. Merriley Borell, "Brown Sequard's Organotherapy and Its Appearance in America at the End of the Nineteenth Century," *Bulletin of the History of Medicine,* 1976, *50*: 309–320; idem, "Organotherapy, British Physiology and Discovery of the Internal Secretions," *J. Hist. Biol.,* 1976, *9*: 235–268; and idem, "Organotherapy and the Emergence of Reproductive Endocrinology," *J. Hist. Biol.,* 1985, *18*: 1–30. Borell examines the role of biologists in the birth control movement in "Biologists and the Promotion of Birth Control Research, 1918–1938," *J. Hist. Biol.,* 1987, *20*: 51–88. On endocrinology and sex research, see Diana Long Hall, "Biology, Sex Hormones, and Sexism in the 1920s," *Phil. Forum,* 1973–1974, *5*: 81–96; and Diana E. Long, "Physiological Identity of American Sex Researchers between the Two World Wars," in Geison, *Physiology in the American Context,* pp. 263–278. Also see U. S. Department of Health, Education, and Welfare, *Today's Medicine, Tomorrow's Science: Essays on Paths of Discovery in the Biomedical Sciences,* prepared for the National Cancer Institute by Judith P. Swazey and Karen Reeds (Washington, D. C.: Government Printing Office, 1978).

15. L. J. Harris, "The Discovery of Vitamins," in Joseph H. Needham, ed., *The Chemistry of Life: Lectures on the History of Biochemistry* (Cambridge: Cambridge University Press, 1970), pp. 156–170; Aaron J. Ihde and Stanley L. Baker, "Conflict of Concepts in Early Vitamin Studies," *J. Hist. Biol.,* 1971, *4*: 1–33; and more recently Naomi Aronson, "The Discovery of Resistance: Historical Accounts and Scientific Careers," *Isis,* 1986, *77*: 630–646. The role of vitamins in nutrition is discussed in E. V. McCollum, *A History of Nutrition* (Boston: Houghton Mifflin, 1957); O. E. Anderson, *The Health of a Nation: Harvey O. Wiley and the Fight for Pure Food* (Chicago: University of Chicago Press, 1958); and Harvey A. Levenstein, *Revolution at the Table: The Transformation of the American Diet* (New York: Oxford University Press, 1988), esp. pp. 147–160.

16. The work of Hans Spemann and Johannes Holtfreter is discussed in Victor

Hamburger, *The Heritage of Experimental Embryology* (New York: Oxford University Press, 1988). Other aspects of the history of embryology, including the increasing importance of biochemical research, are examined in Jean Brachet, "Early Interactions between Embryology and Biochemistry," in T. J. Horder, J. A. Witkowski, and C. C. Wylie, eds., *A History of Embryology* (Cambridge: Cambridge University Press, 1986), pp. 246–259.

17. Joel B. Hagen, "Organism and Environment: Frederic Clements's Vision of a Unified Physiological Ecology," in Rainger, Benson, and Maienschein, *American Development of Biology,* pp. 257–280. Sharon E. Kingsland, *Modeling Nature: Episodes in the History of Population Ecology* (Chicago: University of Chicago Press, 1985); and idem, "Mathematical Figments, Biological Facts: Population Ecology in the Thirties," *J. Hist Biol.,* 1986, *19*: 235–256.

18. Ernst Mayr and William B. Provine, eds., *The Evolutionary Synthesis: Perspectives on the Unification of Biology* (Cambridge, Mass.: Harvard University Press, 1980); William B. Provine, *The Origins of Theoretical Population Genetics* (Chicago: University of Chicago Press, 1971); idem, *Sewall Wright and Evolutionary Biology* (Chicago: University of Chicago Press, 1986).

19. On the role of physicists in molecular biology, see the references in notes 7–9; and Donald Fleming, "Emigré Physicists and the Biological Revolution," in Donald Fleming and Bernard Bailyn, eds., *The Intellectual Migration: Europe and America, 1930–1960* (Cambridge, Mass.: Harvard University Press, 1969), pp. 152–189. The significance of new instrumentation on molecular biology is discussed in Olby, *The Path to the Double Helix,* and Judson, *The Eighth Day of Creation.* For the impact of such instrumentation in biochemistry, see Robert E. Kohler, "Rudolf Schoenheimer, Isotopic Tracers, and Biochemistry in the 1930s," *Historical Studies in the Physical Sciences,* 1977, *8*: 257–298. The impact of instrumentation in physiology and medicine is discussed in Merriley Borell, "Instruments and an Independent Physiology: The Harvard Physiological Laboratory, 1871–1906," in Geison, *Physiology in the American Context,* pp. 293–321; Louise H. Marshall, "Instruments, Techniques and Social Units in American Neurophysiology, 1870–1950," in Geison, *Physiology in the American Context,* pp. 351–369; Frederic L. Holmes, "The Formation of the Munich School of Metabolism," in William Coleman and Frederic L. Holmes, eds., *The Investigative Enterprise: Experimental Physiology in Nineteenth-Century Medicine* (Berkeley: University of California Press, 1988), pp. 179–210; Robert G. Frank, Jr., "The Telltale Heart: Physiological Instruments, Graphic Methods, and Clinical Hopes, 1854–1915," in Coleman and Holmes, *The Investigative Enterprise,* pp. 211–294; and Timothy Lenoir, "Models and Instruments in the Development of Electrophysiology, 1845–1912," *Historical Studies in the Physical and Biological Sciences,* 1986, *17*: 1–54.

20. On developments in embryology and evolutionary theory, see above, notes 16, 18. Donna Haraway analyzes the impact of cybernetics in "The High Cost of Information in Post World War II Evolutionary Biology: Ergonomics, Semiotics, and the Sociobiology of Communication Systems," *Phil. Forum,* 1982–1983, *13*: 244–278; and "Signs of Dominance: From a Physiology to a Cybernetics of Primate Society: C. R. Carpenter, 1930–1970," *Studies in History of Biology,* 1983, *6*: 129–219.

21. William Keith Brooks, *The Foundations of Zoology* (New York: Macmillan,

1899); and William Emerson Ritter, *The Higher Usefulness of Science and Other Essays* (Boston: R. D. Badger, 1918). Brooks's ideas on the purposefulness of scientific research are referred to in M. V. Edds, Jr., "Brooks, William Keith," *Dictionary of Scientific Biography* (New York: Scribner's, 1970), pp. 501–502. On Charles Otis Whitman, see Pauly, "Summer Resort and Scientific Discipline," p. 129.

22. On the role of biochemistry and physiology in medicine, see the essays in Coleman and Holmes, *The Investigative Enterprise,* and the following essays in Geison, *Physiology in the American Context:* Russell C. Maulitz, "Pathologists, Clinicians and the Role of Pathophysiology," pp. 209–235; Richard Gillespie, "Industrial Fatigue and the Discipline of Physiology," pp. 237–262; and Joel D. Howell, "Cardiac Physiology and Clinical Medicine? Two Case Studies," pp. 279–292. Also Frederic L. Holmes, *Claude Bernard and Animal Chemistry: The Emergence of a Science* (Cambridge, Mass.: Harvard University Press, 1974); and Gerald L. Geison, *Michael Foster and the Cambridge School of Physiology: The Scientific Enterprise in Late Victorian Society* (Princeton: Princeton University Press, 1978).

23. Diane B. Paul and Barbara A. Kimmelman, "Mendel in America: Theory and Practice, 1900–1919," in Rainger, Benson, and Maienschein, *American Development of Biology,* pp. 281–310.

24. Thomas R. Dunlap, *Saving America's Wildlife* (Princeton: Princeton University Press, 1988).

25. Borell, "Biologists and Birth Control." On eugenics, see Kenneth Ludmerer, *Genetics and American Society: A Historical Appraisal* (Baltimore: Johns Hopkins University Press, 1972); and Daniel J. Kevles, *In the Name of Eugenics: Genetics and the Uses of Human Heredity* (New York: Knopf, 1985).

26. Emily S. Rosenberg, *Spreading the American Dream: American Economic and Cultural Expansion, 1890–1945* (New York: Hill and Wang, 1982). On the faith in technology, see William E. Akin, *Technocracy and the American Dream: The Technocratic Movement, 1900–1941* (Berkeley: University of California Press, 1977); Joseph J. Corn, ed., *Imagining Tomorrow: History, Technology, and the American Future* (Cambridge, Mass.: MIT Press, 1986); and Howard P. Segal, *Technological Utopianism in American Culture* (Chicago: University of Chicago Press, 1985).

27. Warren I. Susman, "The Thirties," in Stanley Coben and Lorman Ratner, eds., *The Development of an American Culture,* 2nd ed. (New York: St. Martin's, 1983), esp. pp. 225–226. On Robert Millikan, see Robert Kargon, *The Rise of Robert Millikan: Portrait of a Life in American Science* (Ithaca: Cornell University Press, 1982); and Ronald C. Tobey, *The American Ideology of National Science, 1919–1930* (Pittsburgh: University of Pittsburgh Press, 1971), pp. 137–159, 167–198. On Henry Fairfield Osborn, see Ronald Rainger, *An Agenda for Antiquity; Henry Fairfield Osborn and Vertebrate Paleontology at the American Museum of Natural History, 1890–1935* (Tuscaloosa: University of Alabama Press, 1991), esp. chaps. 6 and 7.

28. Daniel J. Kevles, *The Physicists: The History of a Scientific Community in Modern America* (New York: Knopf, 1978).

29. Martin Bulmer and Joan Bulmer, "Philanthropy and Social Science in the 1920s: Beardsley Ruml and the Laura Spelman Rockefeller Memorial," *Minerva,*

1981, *19*: 347–408, on p. 348. Much the same argument is put forth in Roger L. Geiger, *To Advance Knowledge: The Growth of American Research Universities, 1900–1940* (New York: Oxford University Press, 1986); and in Robert E. Kohler, "A Policy for the Advancement of Science: The Rockefeller Foundation, 1924–29," *Minerva,* 1978, *16*: 480–513; idem, "Science and Philanthropy: Wickliffe Rose and the International Education Board," *Minerva,* 1985, *23*: 75–95; and idem, "Science, Foundations, and American Universities in the 1920s," *Osiris,* 1987, *2nd ser. 3:* 135–164.

30. Richard E. Brown, *Rockefeller Medicine Men: Medicine and Capitalism in America* (Berkeley: University of California Press, 1979); John Ettling, *The Germ of Laziness: Rockefeller Philanthropy and Public Health in the New South* (Cambridge, Mass.: Harvard University Press, 1981). The various Rockefeller programs are discussed in Kohler, "A Policy for Science"; idem, "Science and Philanthropy"; Bulmer and Bulmer, "Philanthropy and Social Science"; and in Fosdick, *The Rockefeller Foundation;* and idem, *Adventure in Giving: The Story of the General Education Board* (New York: Harper, 1962). The quotation is from Kohler, "The Management of Science," p. 279.

31. Paul Weiss quoted in Haraway, "Signs of Dominance," p. 134.

32. Gregg Mitman, "Evolution as Gospel: William Patten, the Language of Democracy, and the Great War," *Isis*, 1990, *81*: 446–463; idem, "From the Population to Society: The Cooperative Metaphors of W. C. Allee and A. E. Emerson," *J. Hist. Biol.,* 1988, *21*: 173–194; Cross and Albury, "Cannon, Henderson, and the Organic Analogy."

33. Peter J. Taylor, "Technocratic Optimism, H. T. Odum, and the Partial Transformation of Ecological Metaphor after World War II," *J. Hist. Biol.,* 1988, *22*: 213–244; Haraway, "The High Cost of Information"; idem, "Signs of Dominance"; idem, "The Biological Enterprise: Sex, Mind, and Profit from Human Engineering to Sociobiology," *Radical History Review,* 1979, *20*: 206–237; and idem, *Primate Visions: Gender, Race and Nature in the World of Modern Science* (New York: Routledge, Chapman and Hall, 1989). On the impact of engineering concepts more generally, see William Graebner, *The Engineering of Consent: Democracy and Authority in America* (Urbana: University of Illinois Press, 1987).

34. For example, Bulmer and Bulmer take to task the interpretation advanced by Brown, *Rockefeller Medicine Men,* and the essays in Robert F. Arnove, ed., *Philanthropy and Cultural Imperialism* (Boston: G. K. Hall, 1980). Bulmer and Bulmer, "Philanthropy and Social Science," p. 401, esp. n. 183. In the same vein Ettling, *The Germ of Laziness,* pp. 203–208, criticizes the interpretations in Brown, *Rockefeller Medicine Men* and Howard S. Berliner, "Philanthropic Foundations and Scientific Medicine" (Sc.D. dissertation, Johns Hopkins University, 1977).

35. The Rockefeller work on human biology is referred to in the essays by Kohler and more specifically in the essays by Diane Paul and Garland E. Allen in this volume. On the Rockefeller support for German research on biology and medicine in the 1920s, see Paul Weindling, "The Rockefeller Foundation and German Biomedical Sciences, 1920–40: From Educational Philanthropy to International Science Policy," in Nicolaas A. Rupke, ed., *Science Politics and the Public Good: Essays in Honor of Margaret Gowing* (London: Macmillan, 1988), pp. 119–140.

36. The argument that a core of problems emphasizing evolution, inheritance, and development was central to late nineteenth-century biology is advanced in the Introduction to Rainger, Benson, and Maienschein, *The American Development of Biology,* pp. 3–11; and in Jane Maienschein, *Defining Biology: Lectures from the 1890s* (Cambridge, Mass.: Harvard University Press, 1986), pp. 1–50.

Jane Maienschein

2

Cytology in 1924: Expansion and Collaboration

Cytology had a bumper year in 1924, with two major books published and a third well on its way into print. Together they represented the current state of the science. But they also reveal something more. A close comparison of the second and third of these volumes shows how cytology was changing, expanding into new areas, and becoming more diverse in a way that made it difficult for any one person to handle the entire subject. This expansion of the subject and of its diverse methods of attack reflects the general expansion of knowledge, methods, and enthusiasm within biology. But it also reflects the fact that cytology in particular had reached a stage where what had been a coherent and vital field of study for more than a half century had grown so large and so diverse that it had begun to experience fragmentation and specialization. The three important textbooks that appeared in 1924 and 1925 reflect the status of the field of cell studies as well as the state of biology as a whole.

The first volume was Leonard Doncaster's second edition of *An Introduction to the Study of Cytology*. This leading British cytologist had actually died shortly after the first edition in 1920, but the work had nonetheless undergone revision and updating by assistants. As Doncaster explained in his introduction, he had not intended to provide a textbook in the usual sense. His volume was not designed to summarize known facts and offer a few inferences from them while avoiding any sustained theoretical discussions. That was the purpose of textbooks, he believed, but cytologists were not yet sufficiently unified in their interpretations to warrant such a standard text. Instead, he sought with his volume to "interest the student in the subject by pointing out some of the ways in which cytological investigation is related to the great fundamental problems that lie at the root of all biological research."[1] The organization of Doncaster's book followed his lecture series presented at Cambridge University over the course of six years, to students not previously familiar with cytology. Its chapters considered the basic biological subjects: cell definition, cell organs, cell division, the centrosome, germ cells, fertilization, parthenogenesis, sex determination, chromosomes, and heredity.

At the end of his volume, in a concluding section on the state of the field, Doncaster noted that cytology found itself in a tenuous state just as zoology had after the publication of Darwin's *Origin*. At that time, as he saw it, zoology and physiology had diverged, each pursuing separate questions. He thought that this had been to the detriment of both. Cytology now stood in a similarly precarious position, so that "if cytology is to avoid a similar misfortune, its students must keep in view the need for both the descriptive and comparative and the experimental methods, and remember that the biochemist and physicist are studying with flask and test-tube the same problems that they themselves are attacking with microscope and microtome."[2] Otherwise cytology might also fragment into diverse directions.

The second volume, Edmund V. Cowdry's *General Cytology,* agreed. By bringing together a collection of contributions by American researchers in various aspects of cytology, the edited volume sought to discuss work using both descriptive and experimental methods and work from biochemistry and physics and microscopic study, as well as examinations of both cellular structure and cellular function. The volume thereby attempted to do precisely what Doncaster urged, to keep in view the way that cytology joined different approaches and different perspectives.

Whereas Doncaster believed that a proper textbook in cytology remained premature because of the existence of so many interpretations, Cowdry's group proposed to turn the lack of a single unified view to advantage. By presenting a diversity of facts, interpretations, and methods, the volume could provide a working textbook for general cytology. Its intended audience included students of both biology and medicine, anyone concerned with the nature of the cell.

As Cowdry explained, several individuals had met at the Marine Biological Laboratory (MBL) in the summer of 1922 to begin their collective project. They had wanted to address what was known of cellular structure and function and to consider whether to attempt a cooperative study of the subjects. In particular, given the increase in information and ideas available, they debated whether they should try "to present briefly for the first time within the scope of a single volume data concerning the cell." These data and their discussions would be "fundamental, alike, to the sciences of botany, zoology, physiology, and pathology," so they decided to pursue the project and to coordinate a volume.[3] With such different approaches, and with the different background traditions from which they came, it seemed difficult to bring all the work together in one coherent volume. Some researchers emphasized cell structure using traditional morphological methods of description and comparison. Others pursued the physicochemical makeup and actions of the cells. Some stressed colloids, others chromosomes. Yet the various studies all concerned the same cells, the same fundamental units of life, and as a result a coordinated study seemed well worthwhile. Since no one person was able to cover the full range of ideas and results, the group decided to work together.

They found, Cowdry explained, that the subject "naturally fell into subdivisions." He did not also claim that these represented natural subdivision, or, in other words, that this was in any sense *the* proper way to divide up the subject of cytology. Perhaps there could have been alternative divisions instead, but this set made sense and had the advantage that an investigator at the MBL could handle each part. Each contributor could discuss his or her own line of research and offer his or her own views. This meant that different sorts of data and methods as well as varied interpretations could be included even while preserving the integrity of the whole volume. "In this way the labor involved was shared and did not fall heavily on the shoulders of any single individual. The unique opportunity thus afforded for friendly and informal consultation between the different contributors greatly facilitated the enterprise."

The fact that everyone was together at the MBL and could consult freely and easily made the project possible. And because of that advantage and "in consideration of the fact that several of the contributors had developed their lines of study by availing themselves year after year of the facilities for investigation offered at Woods Hole," the resulting book represented a major MBL effort. The volume was designed to focus on research results rather than on details of methods or historical discussion. It also provided extensive lists of references to direct readers to appropriate further study instead of attempting a comprehensive review of existing literature.

Appropriately enough, Edmund Beecher Wilson provided the introduction. Also appropriately, he devoted the first part of his contribution to some historical considerations, including a review of signal advances up to his own early years of research. Wilson was the grandest figure in cell studies by 1922. Slowed down by severe arthritis, he moved around Woods Hole on crutches, but he persisted with his research and continued to work on the third of the cytology books in 1924, the revision of his own great book, *The Cell*.[4] In his introduction to Cowdry's volume, Wilson noted several factors about the recent past: the growing cooperation between cytology and genetics in studying heredity, the increasing interest in the system of cell components and their action during histogenesis, advances in techniques to study living cells, and—most important—studies of the ways in which altered external conditions affect living cells (including artificial parthenogenesis). Over time, Wilson noted, the result had changed what had been morphological cytology into a new field of cellular biology. This had produced

> a new cytology, a new cell physiology, a new cellular embryology, and a new genetics; and these various lines of inquiry have now become so closely interwoven that they can hardly be disentangled. This much-to-be-desired result has been made possible by an always growing cooperation between lines of attack widely different in method and seemingly in point of view. Such concerted effort in cell research long seemed an almost unattainable ideal; but its realization now seems close at hand. The present book has been undertaken in hope of furthering this cooperation. In the nature of the case it is hardly possible to arrive at complete unity in a work produced by several collaborators representing widely diverse fields of research. Such a group, however, can at least bring to their task a broader and more critical knowledge of the subject than any single writer can at this day hope to command.[5]

In particular, the group as a whole could embrace both physicochemical and morphological perspectives on the cell as no one individual could hope to do. This wide scope is reflected in the selection of chapter subjects and in the overall organization of the volume. Three chapters focus on the physicochemical nature and action of the cell, two on the structure of cells, and five on cellular changes and the role that cells play in various fundamental life processes.

Cowdry's volume provides a marked contrast to the third cytological text. Wilson's own book was undergoing final revision and typesetting at the time that Doncaster's and Cowdry's volumes appeared, and was published the next year.[6] This third edition of *The Cell* followed the same basic organization as the earlier editions but also represented a much revised and extensively updated contribution. Whereas the second edition had had 483 pages, for example, the third offered 1,232 pages. And the general literature list had expanded from twenty pages to fifty-eight pages, though even then Wilson apologized that his list represented only those selected works actually cited in the text and not a comprehensive bibliography.

In his first edition of 1896 and revised second version in 1900, Wilson had sought to survey the new field of cell biology. Thirty years later, following the rediscovery of Mendel and the resulting explosion in the study of heredity, the task of providing a full survey seemed much less reasonable. In addition, the relations of cytology to the other fields of anatomy, histology, embryology, physiology and genetics, which had been recognized in 1896, had expanded to include close relations with cell physiology, biochemistry, and biophysics. This wide range of relations made cell study "so diversified that no single work could possibly cover more than a small portion of it." As a result, "extended general treatises on cellular biology have largely gone out of fashion in favor of more circumscribed works dealing with particular aspects of the subject, and thus making possible a more intensive treatment." Yet despite the difficulties, Wilson still "ventured to think that the need of a work of somewhat more synthetic type has not disappeared."[7]

Wilson's own emphasis remained concentrated on the structure of cells and their roles especially in development and heredity. Zoological cytology and embryology predominated, while physiological or biochemical concerns and botanical study of cells remained less central. Chapters in the second edition on cell division, germ cells, fertilization, chromosome reduction, cell organization, cell chemistry and physiology, cell division and development, and inheritance and development largely remained in the third. Each chapter expanded, most about doubling by adding new information and consideration of new interpretations. In addition, new chapters appeared on reproduction and sexuality and others on chromosomes. These Wilson saw as the major areas of advance calling for inclusion.

Many of the subjects in Cowdry's volume received little attention in Wilson's. Even those included in both books, such as heredity, fertilization, and differentiation, were approached very differently. Where Wilson's book offered a meticulously detailed consideration of as wide a range of existing literature on his subjects as possible, Cowdry's included deeper exploration of several selected sets of facts, methods, and ideas. Where Wilson sought to remain judicious in his consideration of alternative theories, contributors to Cowdry's volume willingly advocated their various theoretical interpretations. While Wilson necessarily provided one particular judgment and one set of biases, the cooperative project brought together a mixture of alternative views. In the face of expanded knowledge, the two works offered interestingly different sorts of synthetic treatments of cell biology.

A Comparison of *General Cytology* and *The Cell*

In order to demonstrate the way in which Cowdry's volume differs from Wilson's, it is useful to adopt a closely descriptive and comparative approach. This will allow a careful look at each subject addressed and an examination of the data discussed, the approach, and conclusions. Since the authors in Cowdry's volume wrote their contributions expressly for this special volume, many of the papers have a somewhat different flavor than the usual professional publications by the same scientists. Here they consciously seek to reach a wider audience in both biology and medicine. Some provide relatively straightforward surveys. Others also bring more general themes and theoretical considerations into the discussion than they typically would have.

The list of contributors to Cowdry's volume includes the MBL leaders, who were also national leaders in these areas of cytology and physicochemical studies of cell processes. All members of the group had their Ph.D. degrees, except for the Lewises. Warren Lewis had received an M.D. from Johns Hopkins, and Margaret Lewis (formerly Margaret Reed) had received her B.A. from Goucher College and had then continued at Bryn Mawr for a year. Only Robert Chambers had gone abroad for his degree, the rest studying at Columbia (Albert Prescott Mathews), Pennsylvania (Merkel H. Jacobs), Chicago (both Frank Rattray and Ralph Lillie, Edmund V. Cowdry, Ernest Everett Just), Johns Hopkins (Edwin Grant Conklin, Thomas Hunt Morgan), and Kansas (Clarence E. McClung). Many had published or were soon to produce major texts in one or another cytological field, and all published extensively.

Aside from Wilson, the oldest of the group were Conklin (at fifty-nine) and Morgan (fifty-six), whereas the youngest was the editor (Cowdry at thirty-four). The others ranged from nearly forty to their early fifties. Most worked at leading research centers around the country, though not all in traditional university settings. For example, Ralph Lillie had moved to the Nela Research Laboratory in Cleveland after having taught at several schools including Pennsylvania and Clark University. Cowdry had taught at the Peking Union Medical College before he returned to the United States and moved to the Rockefeller Institute for Medical Research in 1921; Margaret Lewis worked in her husband's laboratory at Johns Hopkins Medical School. And Ernest Everett Just taught at Howard University rather than at the sort of major research institution he would have preferred.[8]

All were regulars at the MBL. Conklin and Morgan had spent most of their summers there since their graduate school days. Indeed, both had served in nearly all capacities there: teaching courses, giving public lectures, serving as trustees, actively working with fund raising, and introducing new generations of graduate students to the community by taking them along for summers of research activity. Frank Lillie had served as right-hand man to the first director, Charles Otis Whitman, since the 1890s and had become assistant director and then director of the laboratory. As his student, Just had also become part of the MBL group each summer by the 1920s, as had Ralph Lillie. McClung had probably joined the group during the time he visited Columbia to work with Wilson, as had Mathews during his studies at Columbia. Warren and Margaret Lewis were friends of Ross Harrison, another active MBL member and trustee from Johns Hopkins, and they had participated in earlier special programs related to their work on tissue culture and neurobiology. Jacobs became a trustee and then the third MBL director when Lillie retired, serving in that role from 1926 to 1937. Clearly, this group represented something like a "ruling class" at the MBL, and its interest in cytology reflected one of the central interests of the institution.

Cell Chemistry

The first chapter of Cowdry's volume following Wilson's introduction was Albert Prescott Mathews's on cell chemistry. The general chemistry and physiology of the cell was a subject that Wilson felt incompetent to discuss in full and which, as a result, he felt remained to be explored. Yet in his 1925 edition, Wilson looked briefly at colloidal theories of the cell and followed Jacques Loeb's suggestion that the cell operates essentially as a chemical machine that transforms food materials into living and functioning form. In a short half chapter, Wilson considered the available evidence about the chemical nature of the various cell parts, including the nuclear and chromosomal elements. There he offered his only brief direct references to Mathews's work, that on the nature of nucleic acids. Wilson also said little in his book about physics or about the special chemical nature of living cells, though he did insist that *organization* rather than some vital substance distinguishes life from nonlife. In effect, form characterizes living matter, but Wilson did not attempt further to define life.[9]

In marked contrast, Mathews directly tackled questions about the nature of life. He moved beyond solid reporting of data and research methods to enter the exploratory spirit of the volume by giving vent to some interpretive speculations. Mathews had taught a course on cell chemistry at the University of Chicago and presumably had developed his broader picture of the cell in that context. His essay begins by pointing out that physics and chemistry had only recently begun to provide much help for study of the cell. Recent advances had made such an enormous difference that the biochemist could, in fact, begin to make sense of the cell as a machine. Specifically, and here he disagreed with Loeb's chemical emphasis, Mathews believed that it is an electrical machine. This knowledge transformed the biochemist into an engineer, Mathews believed, though an engineer in the process of becoming rather than an accomplished expert since he could take the electrical and mechanical system apart but could not yet put it together again. Making repairs and creating similar living machines must wait.[10]

According to Mathews, to reach these constructive goals requires understanding that "most characteristic" element of living things, namely the "psychic element." Psychism must be part of the biochemistry of the cell, Mathews insisted, for without it the cell would be like *Hamlet* without Hamlet. The psychic component is just as much a part of nature as gravity or inertia, so that the biochemist must be part poet and part psychologist as well as an electrical engineer. This psychic element might well be particulate, he suggested, like matter, light, and energy. Or it might not. Yet he did not intend to suggest any sort of metaphysical dualism, for at its root life, like the physical world, is made up of material substances.

Beyond the psychic, Mathews provided a list of facts that any theory about the chemistry of cells must also explain. For example, the fact that life ceases without oxygen, or, in other words, that living protoplasm must be in a state of partial oxygenation—this is fundamental to understanding life. The ability of cells to grow by synthesizing proteins and other substances is likewise vital. So is the capacity of living things to generate electrical currents. These, then, are the "fundamental phenomena to be explained."[11] And they should be attacked through a coherent look at all the phenomena together within the context of one theory. It was particularly important for Mathews that any theory must correlate all the facts together in order to be considered successful.

Turning to physics in his lengthy essay, Mathews explained that electrons make up life. They are then organized into atoms and, further, into molecules. These bundles of moving electrons make up tiny "universes" which move through space. Yet Mathews argued that this space is not empty, as contemporary physics suggested, but is instead filled by an ether. In describing the ether and its importance, Mathews appealed to Sir Oliver Lodge's popular 1902 volume, *The Ether of Space*.[12] Living things are little universes, Mathews agreed with Lodge, with space filled with the luminferous ether. Science must study the interaction of ether and matter, he insisted, not only to understand life but also to understand physics and chemistry. "For life illumines physics and chemistry just as truly as physics and chemistry have illumined physiology and psychology. There is more in matter than meets the eye." Further, ether is space multiplied by time; ether is infinity and eternity; ether consists of particles called etherions; and ether is not a mere metaphysical construct but is a "real physical, as well as perhaps a psychic entity." Within ether, energy is an "etherial flux or motion" and is somehow "the same as mass," both being a rotation in the ether. And molecules exist within the ether, as systems of atoms.[13] Lodge's eloquence, itself inspired by Ernst Haeckel's earlier versions of materialistic monism, clearly moved Mathews to speculative enthusiasm.

Only halfway through the article did Mathews deal with more familiar biological phenomena such as respiration or growth of cells. Only two-thirds of the way through did he attempt to develop the idea of the cell as an electrical machine, or a battery, in a more concrete way.

Chromatin provided the last major subject that Mathews addressed, though it was here that he had done the most significant original research. Although nucleic acid was hard to obtain and thus to study, Mathews outlined in some detail what was known of the chemistry of chromatin. He then considered the chromosomal theory of inheritance, according to which chromosomes are the carriers of all inherited material. Echoing a view he had put forth in his *Physiological Chemistry* of 1915, Mathews rejected the chromosome theory as very improbable, partly because he felt that chromosomes are just too simple in their composition to carry out such a complicated hereditary task as was assigned them. Far from their having proved the theory, Mathews thought that the

> onus of proof is on those who assert that the chromosomes are such museums containing samples of all the chromatin of all the cells of the body, not only all the chromatins which develop during life, but all that infinite collection of old masters inherited from the past, and all the infinite number of descendants yet to appear in the eons before us, and presenting qualities usually said to be dormant. They are concealed no doubt in the chromosomal attic, ready to be produced when the occasion arises.[14]

Such an idea struck Mathews as unsupported and as probably unsupportable.

Obviously, Mathews's views were not universally held. Wilson, who accepted and had helped to develop the chromosome theory, believed that Mathews had overlooked some basic facts. Another critic found Mathews's speculative ideas rather odd, complaining that they "may mean something to the metaphysician, but one cannot help feeling that Prof. Mathews's view on the relationship between cell lipins [sic] and cell proteins, or on the biochemistry of development, would have been more useful."[15] Yet Mathews was not simply spinning out the sort of crackpot ideas that ended up in Frank Lillie's file cabinet at the University of Chicago under "C.R.A.N.K.S." Although rather temperamental and although involved in various clashes with colleagues both at Chicago and elsewhere, Mathews generally commanded high respect from other biochemical physiologists. His ideas and his approach were unorthodox but always provocative and exciting. He taught physiology at the MBL for seventeen years and remained a major figure, first as a department leader at Chicago and then at Cincinnati after he moved there to head a new physiological chemistry department in 1918.

Mathews belonged to a group of biologists in the early decades of this century who saw the organism as a unit rather than as a straightforwardly reducible set of physical parts and actions. His insistence on psychism as part of cellular phenomena did not obviously conflict with other ideas about design-in-nature by Ralph Lillie and others, or the insistence on the organism-as-a-whole by Charles Manning Child or even Thomas Hunt Morgan in his early career. Some physicists also tended to have similar views, including a few of the physicists who were invited to lecture at the MBL to encourage cross-fertilization of ideas from different disciplines. Although Lodge's popular books of the late 1800s and early 1900s and their particular details of physical theory had become rather outdated by 1924, many would have agreed with the impulse there. Somehow nature *must* be more than a bundle of separate physical pieces, such thinkers suggested. Coordination of parts might take different forms, but whatever the form, science must seek to understand the nature of the unities and the coordination. Such coordination of parts, or "organization" as Wilson called it, is what makes life work, after all.[16]

Epistemically, many scientists sought general theories such as Mathews's that could explain the whole range of facts at hand. A unified theory, even if dependent on rather questionable suggestions, might succeed better according to this view than a more careful but limited and less provocative theory. Thus, stimulating ideas that might prove horrendously wrong in the long run might very well represent first-rate science at the moment, if they provided a sufficiently suggestive framework within which to work. Mathews did provide such suggestions, though, as Robert Kohler puts it, "Not unlike the molecular biologists of a later time, Mathews was regarded by the more sober citizens with a mixture of awe and alarm."[17]

Cell Permeability and Reactions

For Wilson, cells remain individual units even while they interact with other cells to make up a whole organism. As a result, the cell membrane for Wilson serves primarily to define the cell and to separate it from its environment.[18] The cell exists, for Wilson, as a starting point; every individual organism begins as a cell. Thus he did not devote much attention to the way a cell arises or to the special role of the cell membrane in regulating the substances that are allowed to enter or those excluded. Questions about permeability of the cell membrane did not seem as central a question to Wilson as it did to others.

Merkel H. Jacobs, for example, saw the differential permeability of the cell to different substances as basic to cytology, making cells what they are. Life is dependent on the ability of cells to regulate which substances reside inside the cell and which are kept outside. This often complex regulation produces the internal heterogeneity of material necessary for life.[19] Many of the most central questions about the nature of body functions depend on the results of differential cell permeabilities.

To study permeability, Jacobs believed, the researcher had best adopt the widest range of diverse methods available. With each method, it is difficult to know whether experimental conditions remain close enough to normal conditions to provide useful information. In addition, slight differences in conditions in each case might produce slightly different and confusing results. Therefore he warned that the researcher must remain particularly careful not to generalize from single cases.

After reviewing the various alternative theories about what causes differential permeability, Jacobs insisted that no one theory had gained significant credibility as yet. It remained to generate more facts, in particular by covering a wide range of materials and by using as many different methods as possible. Convergence and agreement would strengthen the results of each separate study. Finally, examining what sorts of factors can bring changes in cell permeability under experimental conditions would also provide information about what directs normal permeability. Then, according to Jacobs, "a satisfactory theory will follow as a matter of course. Until that time, speculations should be reduced to a minimum."[20] Wilson would have agreed with such an epistemological preference even while he focused on different questions, though Mathews would not have.

In looking at cell reactions, Ralph Lillie, like Jacobs, took on a more defined topic than Mathews had. Yet, like Mathews, Lillie (who was also at the University of Chicago) was interested in what makes the whole cell and the whole organism of coordinated cells work. Like Jacobs, Lillie saw the cell as essentially interconnected with its chemical environment and with other cells. All concentrated on chemistry and physiology much more than did Wilson, who maintained his morphological focus even when he acknowledged the interrelations of cells.[21]

For Lillie, the cell is not an enclosed, isolated thing, but it lives in equilibrium with a changing environment. Lillie asked how the cells react, or adjust their functions, in response to the varying conditions in that environment. In other words: "What are these special features in the composition or constitution of living matter which render its chemical processes so susceptible to influence by changes in the surroundings?"[22]

Basically, Lillie held, the cell undergoes changes in metabolism when a stimulus acts on the protoplasm and modifies the chemical reactions, in part by affecting the reaction rate which depends ultimately on the particular structure of the protoplasm. Therefore Lillie proposed to begin by looking at studies of experimentally altered structures and the effects of changed external conditions, though he thought that studies with nonliving material could have only limited value for illuminating ordinary living processes.

Another line of research suggested to Lillie that protoplasm is like an emulsion of oil drops in water: it consists of material in two different fluid states. Stability between the two depends on the presence of thin films which act as surface layers. And research on emulsions showed that these layers can be broken down or established quite easily, with only a small change in ion concentration or the presence or absence of a tiny amount of a particular substance making all the difference between an emulsion and a layering of two different materials. For Lillie, the living cell seemed to be a proper emulsion, with a thin film in the form of the outer cell membrane keeping the cell intact and separate from its surroundings. Evidence had also accumulated that the material inside the cell is also divided by thin film partitions between chemically different parts. The semipermeable nature of the partitions accounts for the sophisticated regulations within the cell.

Yet the films do not provide a permanent or irreversible partitioning of the cell. Their semipermeability allows diffusion of substances across the films and makes regulation possible. And even a small stimulus can have a far greater effect than might seem possible. Thus a tiny pinprick in one spot may affect the entire cell in a radical way. It seems, Lillie concluded, that the cell has some sort of transmission process to carry effects throughout the whole. The propagated effect apparently travels along defined paths, in a manner similar to that of a neuromuscular response: a stimulus in one place on the nerve can travel rapidly along the nerve and can then stimulate the reaction of muscle fibers or other reflex actions. The whole cell must regulate its response to remain in proper equilibrium with its surroundings and within itself.

All the facts seemed to favor one theory over others, namely that electrical stimuli control all the stimulation and transmission of cell reactions. Presumably the current causes polarization along the boundary of the films, with resulting chemical changes along the film. These are transmitted beyond to the rest of the protoplasmic material. A survey of various cell and organismal functions suggested that all fit within the electrical transmission theory, which therefore provided a unified way of looking at the organism as a whole as well as at the individual parts.

Wilson referred in a very positive way to Lillie's earlier study of permeability, even acknowledging that the changes in permeability of the entire cell that occur at fertilization had been well documented. Furthermore, he agreed that the process might involve a changing electrical equilibrium, as Lillie suggested. But he devoted no further attention to that subject since, as he put it, "we are here concerned more particularly with the cytological changes."[23] By this, Wilson meant morphological changes, though Lillie, Jacobs, and Mathews would clearly not have agreed that these were more truly "cytological" than the sorts of actions affecting the whole cell that they studied.

Cell Structure

Cell structure was something Wilson did regard as basic. He was centrally interested in the different special parts of the cell as well as in structure and organization within the familiar general protoplasm. Chapters on the "General Morphology of the Cell" and on "Some Problems of Cell-Organization" most directly consider the subject. Wilson traced a collection of alternative theories about the nature of protoplasm, with some theories stressing the fibrillar or reticular structure, others the gelatinous viscous nature, for example. He concluded that no one theory had yet gained general acceptance. Perhaps there is a basic invisible structure, which admittedly may force "us back upon the assumption of a 'metastructure' in protoplasm that lies beyond the present limits of microscopical vision; but in that respect the biologist is perhaps in no worse case than the chemist or the physicist."[24]

In his essay, Robert Chambers agreed with Wilson. Both maintained that the protoplasm must have a defined structure. Furthermore, it must be more than a liquid since the protoplasm serves as the center of all life and appears to be highly organized. And it must be more than a mere random collection of protoplasmic material. The cellular unit must have its protoplasm, its nucleus, its cortex, and other parts, and some organization. Therefore it "must be regarded not as a 'stuff' but as a mechanism consisting of visibly differentiated and essentially interrelated parts."[25]

The problem was to devise a way to see the structure, much of which may remain invisible even with advanced microscopy, since the protoplasm normally resides inside the cell and hence remains unobservable under normal conditions. People had tried crushing the cell to determine its viscosity. Or centrifuging it to assess the effects of displacement under a change in gravitational force, or electromagnetic experiments for parallel reasons. Or microdissection with tiny needles, using tissue suspended in hanging drops. Or even microinjection of substances to determine the effects on internal structure. None of these methods had proved perfect. None gave a definitive answer, for example, to the question whether the different areas of the cell remain structurally or functionally independent or whether they are connected through a larger reticulum.

As Chambers pointed out, the evidence had accumulated to suggest that the cell has defined and stable different parts. The nucleus maintains its integrity because of its membrane, which if disrupted releases the nuclear substance throughout the cell. Furthermore, the destruction of the nucleus leads to a disruption of cell function as it becomes unable to divide. Chambers spent considerable time discussing the nucleus and the structural changes of all its parts, including the chromosomes, during cell division. Much of this work Wilson also referred to and accepted in *The Cell*. But in his essay Chambers offered a very different emphasis, namely on the methodology of microdissection and injection more than on the results. Like Wilson and unlike Mathews, Jacobs, and Lillie, Chambers also retained a largely morphological approach to cellular protoplasm.

Like Chambers, Edmund Cowdry pursued a largely morphological topic in his essay. Yet he also sought to show the connections of structure to function of cellular parts, especially the mitochondria and Golgi-apparatus. In particular, he wanted to show how the chemical nature of cellular parts dictated their functions. Unfortunately, "our functional interpretations must necessarily lag far behind on account of the great difficulty of projecting accurate methods of chemical analysis into such very small units as cells."[26]

Cowdry lamented that researchers had tended to generate more and more data using the same observational methods. Instead, they needed new alternative experimental approaches to gather different sorts of data as well. For Cowdry held that looking at the broader picture and gathering a variety of evidence would make a stronger case:

> It is possible that we, as students of mitochondria, have allowed ourselves to become rather narrow, and have approached too closely to the problem to see it in its proper proportions, whereas our task is really a synthetic one: we must piece together information from many quarters, and build up in our mind's eye a dynamic picture of mitochondria in relation to innumerable other cellular constituents. To take a familiar example, the close study of the mainspring of a watch would not tell us much unless its behavior was carefully considered in connection with all the other parts of the mechanism.[27]

Mitochondria clearly are present in almost all the major vital phenomena, Cowdry saw, which implies that they play a central role, perhaps in respiration. But too little was known to draw any safe conclusions about mitochondrial function as yet.

In this, Cowdry and Wilson largely agreed. Wilson pointed out that a likely role offered for mitochrondria in fertilization remained completely speculative. One full-scale theory suggested that mitochondria enter the egg from the sperm at fertilization, are passed on to other cells after division, that they then give rise to more specialized protoplasmic structures, and that throughout this cycle they retain their integrity and also play vital roles in heredity and development. Wilson joined Chambers in acknowledging that parts of the theory had "far outrun the facts" or even faced contradictions with observation, but that nonetheless the emphasis on mitochrondria "has too many facts in its favor to be lightly dismissed."[28]

Likewise, Wilson felt that "too little is known of the Golgi-apparatus, morphologically and physiologically, to warrant extended discussion at this time." It was not even clear whether they retain identity for long, though some evidence suggested that they did. Nor was their function at all clear. In fact, Wilson believed that the basic studies of Golgi-bodies were, like those of mitochrondria, "still in a somewhat unformed state."[29]

Cowdry agreed. First, the Golgi-apparatus did not really appear to him to be a distinct organized "body" but rather a collection of chemical substances. Neither microscopic work with vital stains and living cells nor study of prepared materials showed the actions or exact structure of the Golgi apparatus. This left the suggestion that here was another functioning part of the cell, but the cytologist had little to work with to date.

The so-called "chromidial substances" posed similar problems for both Cowdry and Wilson. There seemed to be defined, observable (since they stained regularly) substances in the nucleus. These apparently interacted with materials in the remaining cytoplasm. Yet the exact structures of the materials and the precise nature of the interactions remained unknown even though more was known of structure than of function. Wilson and Cowdry shared the view that here was a potentially important subject warranting careful attention with both traditional morphological and newer experimental methods. And they agreed that little was yet certain. Cowdry's attempt to look at cell parts beyond the long-recognized chromosomes and other nuclear parts remained more suggestive than final. Yet it received special attention when one reviewer of the volume applauded the fact that "the much-abused Golgi apparatus has at last received official recognition."[30]

Cell Behavior

Most of Wilson's book is dedicated to the structure and activity of cells under normal conditions. Experimental manipulations provide additional useful information toward understanding the normal cases, certainly, but Wilson focused on descriptions of the normal rather than on discussions of experimental conditions or results. The Lewises' paper in Cowdry's volume followed much of the rest of their work in exploring what happens in experimental cases.[31] Specifically, what do various types of cells do and what structures do they have in various sorts of tissue cultures? Tissue cultures allow the researcher to study individual functioning cells or cell clusters, which normally would be hidden away well inside the living whole organism.

Tissue culture is possible because cells remain "alive" for some time even after the host organism is dead or after the cells have been removed from their initial living whole. Hanging drops of appropriate fluids provide useful culture media in many cases and offer the advantage that the researcher can easily observe the results. When successful, cells will grow and divide within the medium, then move outside as well. This makes it possible to observe changes involved in cell division, reproduction, and other crucial activities of life. As Warren and Margaret Lewis showed, some cultures can be maintained for long periods, undergoing division after division, and promising potential immortality of sorts. Some cells experience differentiation within the cultured tissue. Others undergo "dedifferentiation" as the cells begin to die when the culture medium is not renewed sufficiently or often enough; in that case renewal requires adding embryonic extracts from living materials rather than inorganic materials. As the Lewises said, such phenomena remained difficult to interpret.[32]

The major question centered on the extent to which cells in tissue cultures act normally and therefore reliably reflect normal activities. The Lewises' essay implies that the results are useful and virtually normal. Wilson agreed, despite persistent arguments to the contrary. In particular, many biologists insisted on the integrity of the organism as a whole. Indeed, Wilson acknowledged the importance of integration of the parts in a normal complex organism, in which each cell remains just a part within a whole. Yet for Wilson the composite and coordinated whole comes secondarily. Fundamentally, each cell "possesses in itself the complete apparatus of life." And "we shall therefore proceed upon the assumption, if only as a practical method, that the multicellular organism in general is comparable, to an assemblage of Protista which have undergone a high degree of integration and differentiation so as to constitute essentially a cell-state."[33]

Fertilization

The next chapter in Cowdry's volume turned to the behavior of cells in fertilization. With Frank Lillie as senior author, and with his name placed out of alphabetical order ahead of his coauthor, E. E. Just, many people naturally assumed that Lillie was the one who had written the bulk of their essay on fertilization. Yet Just's biographer, Kenneth R. Manning, suggests that in fact Just wrote at least an equal share.[34] Just had carried out his graduate work at the University of Chicago and at the MBL under Lillie from 1909 to 1916, when he finally found enough time away from his teaching position at Howard University to complete the Ph.D. During that time he had pursued the phenomenon of fertilization and was strongly influenced by Lillie's theory of the importance of a special substance which Lillie called fertilizin. After publishing several papers on various details of the fertilization process, Just turned with Lillie in this article to a defense of Lillie's point of view—with reservations. Perhaps the joint authorship allowed Just to explore some of these reservations in a friendly context.

Fertilization actually includes all stages of the union of ripe gametes and not only the stages after the union is effected, according to Lillie and Just. This phenomenon, in which two cells become fused into one, is unique and happens nowhere else in nature. The very first changes are those along the cortex, as a "fertilization membrane" forms and a "wave of negativity" (which prevents further sperm penetration of the egg) results. Both egg and sperm take active roles in the process, they said, and it is not true that the sperm simply bores into the passive egg as many had held. For example, German cytologist Theodor Boveri had suggested that the sperm carries a centrosome into the egg, which acts as the "active division center" and which then initiates all cell division. But such a theory no longer fit the facts in the 1920s, Lillie and Just maintained.

Something about the fertilization process also produces a substance that causes agglutination of the remaining sperm cells. As such, agglutination is a part of fertilization and might provide information about the normal union of one sperm with the egg. Jacques Loeb had suggested that the dissolving of the jelly coat around the egg, which occurs upon fertilization, causes agglutination. Yet Lillie's and Just's experiments suggested otherwise, though the exact nature of the chemical reaction remained unknown. Yet it was clear that Just and Lillie believed Loeb to be wrong in his views.

Historians have begun to look at the rivalry underlying this discussion of Loeb's ideas.[35] Lillie and Loeb had very different approaches to science, as to life, and perhaps were bound to clash. While both worked at the University of Chicago, Lillie remained a student and junior faculty member under Whitman whereas Loeb headed the physiology program. There Loeb developed his ideas of fertilization with relative success. At the MBL and at Chicago, Loeb discussed his work on normal fertilization and on artificial parthenogenesis. Loeb was in poor health by 1922, however, and he died in 1924, so that his side of the story is not represented in Cowdry's project.

Artificial parthenogenesis raised interesting questions since its very existence suggested that there was something nonspecific about fertilization. If altering the salt concentration in sea water could stimulate cell division, fertilization might not be such a complex or special life phenomenon as it seemed.[36] As Loeb developed the implications of his view, he objected especially to such alternative interpretations as Lillie's theory of fertilizin.

A Philip Pauly puts it, Loeb rejected Lillie's very approach, for "fertilizin was a hypothetical substance of indefinite nature whose complex structure was defined in terms of the event it was asked to explain." Further, Lillie did not appeal to knowledge of physical chemistry, as Loeb preferred, but rather he was "offering words" and "thus incorporating his theory into all descriptive discussion."[37] In putting forth his views in 1916, Loeb had directly attacked Lillie's fertilizin theory, and Just had criticized Loeb in response.[38] This essay continued the criticism, and the dispute surfaces in several places, though generally in rather restrained form. The grounds of dispute were psychological, epistemic, and methodological more than specific questions about facts.

Lillie and Just did point out some strengths of Loeb's chemical emphasis, but they then went on to criticize his cytolysis theory of parthenogenesis and fertilization, according to which fertilization allows a cytolytic agent (which Loeb had called lysin) to break down the egg cortex. For many reasons, "we cannot admit that Loeb's conception, though it was a powerful stimulus to research, contains a workable hypothesis of activities."[39] Although they suggested that "no single theory can account for all the phenomena of fertilization as we have defined it," in fact the essay serves as a defense of Lillie's theory and approach.

Wilson did not wholly adopt either Lillie's or Loeb's interpretation. Rather he accepted and rejected parts of both. Instead of developing a theory about exactly what happens in fertilization or what chemical changes occur, for example, Wilson gathered a wide range of evidence from different researchers. As he did throughout his volume, he reported the results of many experimental studies, laid out various alternative theories, discussed the theories in light of the evidence, then offered some tentative conclusions.

The fact that Wilson's chapter on fertilization also included parthenogenesis reflects the importance he assigned it. Artificial parthenogenesis could be made to initiate a sequence of events that very closely paralleled normal events in the egg. "To the cytologist," Wilson concluded, "the processes called forth by fertilization or parthenogenetic activation offer the appearance of a single train of connected events, more or less plastic in each individual case and varying materially in its details from species to species."[40] Apparently, because of heredity and organization, each egg is capable of dividing and differentiating to some extent. What was needed, Wilson saw, was further experimental study of precisely what physiological changes take place in fertilization and subsequent differentiation to determine the normal conditions.

Differentiation

Another aspect of cell behavior was differentiation of individual cells within the whole organism. In his essay in Cowdry's book, Wilson's long-time friend Edwin Grant Conklin addressed the classic question: how do individual cells undergo differentiation from a "more general and homogeneous to a more special and heterogeneous condition?"[41] Protoplasm goes through cycles of differentiation, he suggested, then cycles of dedifferentiation. Yet there is no such thing as undifferentiated protoplasm, for every cell is differentiated into parts from the beginning: nucleus, cytoplasm, centrosomes, aster, sphere, and so on. The life cycles of individual organisms exhibit development patterns of "progressive differentiation" combined with integration to effect the whole. The differentiation transforms the general material into specialized structures and functions.

Differentiation arises, Conklin explained, through epigenetic processes and certainly not through any process of qualitative division of predifferentiated parts. He directly rejected the sort of predeterminist interpretation of development and differentiation that Wilhelm Roux and August Weismann had proffered in the 1880s and 1890s. Nuclear division does not serve to separate out particulate inherited determinates into different cells, which become structurally and functionally differentiated simply in accordance with that differential inherited information.

The details of cell lineage demonstrate this fact, Conklin said. Tracing the exact fate of each cell and the pattern of each cell division through many cleavage stages shows that some cleavages are determinate and others are indeterminate, for example. This means that some organisms and some cells do not exhibit the regularities in cleavage that others do. Where regularity and determinate division occurs, there must be some underlying "structural peculiarity of the protoplasm," Conklin concluded. The divisions serve to isolate different materials into different cells.[42] In these cases, the cytoplasm thus largely directs differentiation. Yet in determinate and indeterminate cases alike, the nucleus and the interaction of cytoplasm and nucleus remain vital as well. Most of what Conklin said in this essay summarized his conclusions from his cell lineage study from 1890 to 1905. After that time, he continued to work on different organisms and on different details of each. But the principles had all been in place by the time of his major 1905 publication on ascidian development.[43]

After 1905, Conklin addressed questions of heredity more directly. Based on a careful reading of other research and on his own results, he decided that a Mendelian interpretation of heredity best fit with the facts. By the time of this essay in 1925, he had long argued for a Mendelian-chromosomal interpretation of heredity and had coordinated it with his cell lineage studies of cytoplasmic development. At the end of this essay, he asserted that there was conclusive evidence in favor of: the existence of Mendelian factors (or genes) on chromosomes, the halving of chromosomes during mitosis, and the resulting similarity of chromosomal makeup in every cell of the same body. But one major question remained for Conklin: given these facts, how can we explain how identical genes correlate with differentiated cells?

We must look at the whole cell, he insisted, for the cytoplasm holds the answer. Recall that Conklin stressed the interaction of nucleus and cytoplasm. Yet the nucleus need not direct the cytoplasm. Indeed, the evidence lay in the other direction.

> Differential cell division is the result of definite movements of the cytoplasm, of definite orientations of spindles and cleavage planes, and ultimately of a definite polarity, symmetry, and pattern of the cytoplasm. There is good evidence that these movements, orientations, and localizations in the egg are the immediate results of cytoplasmic activity; these activities may themselves be the results of the interaction of nucleus and cytoplasm at an earlier stage, and possibly the inherited differential for all these orientations of development may be found in chromosomes or genes.

In short, "some of the differential factors of development lie outside of the nucleus, and if they are inherited, as most of these early differentiations are, they must lie in the cytoplasm."[44] Conklin was not about to become a friend of the nucleus alone—or of the cytoplasm—but remained an exponent of cytoplasmic as well as nuclear direction of development and inheritance.

Since Wilson's work had closely paralleled Conklin's, he had a great deal more to say about differentiation than about such physiological concerns as the early essays in Cowdry's volume had discussed. Differentiation appeared as a section in Wilson's lengthy chapter on "Development and Heredity," the final and conceptually central section of his book. There Wilson explained that though the egg operates as a "reaction-system" which responds to external conditions, "the specific differences of development shown by these various animals must be determined primarily by internal factors inherent in the egg-organization." Heredity, or the "innate capacity of the organism to develop ancestral traits," contributes to the particular organization of each egg. So does development, or "the sum total of the operations by which the germ gives rise to its typical product."[45] The organization remains absolutely central, yet we know little about what causes it. What is clear is that both heredity and development play vital roles in effecting and directing organization.

Differentiation within the organized egg occurs at all stages: before, during, and after each cell division. To date, the "existing knowledge of this subject is still too fragmentary and discordant to offer a sufficient basis for adequate discussion," Wilson insisted. The problem was to discover the mechanics of localization and differentiation and to determine what role heredity plays and how. Whereas Conklin had insisted on the dual importance of cytoplasm and nucleus, Wilson left open the question of exactly how the mechanisms of differentiation work. Chromosomal interpretations, and specifically Mendelian accounts, had begun to suggest answers to many questions already. Therefore Wilson reserved judgment but adopted a hopeful view that though "we are confronted still with a formidable array of problems not yet solved, we may take courage from the certainty that we shall solve a great number of them in the future, as so many have been in the past."[46]

Theories of Heredity

The last two essays of Cowdry's volume turned to those sources of Wilson's optimism, to the chromosomal and Mendelian theories of heredity. In the first, Clarence E. McClung addressed the chromosomal theory and maintained that it explained a great deal. While working as an advanced student in Wilson's laboratory at Columbia in 1902, McClung had discovered that in some insects (the Orthoptera) all and only the males have an accessory chromosome. At first he had remained cautious in his interpretations of that accessory chromosome, saying that "regarding the theory of its function advanced in this paper, I can say only that it has, if anything, been strengthened by later researches, and more nearly explains the phenomena involved than any other that has been conceived."[47] Despite his hesitations and despite some initial errors in his research, McClung quickly gained clarity of results and confidence in his interpretation. By 1925, he had become a defender of the chromosomal interpretation of heredity.

Yet like Conklin, McClung did not see the chromosomes as sharply distinct cellular bodies that sit there in a superior way giving out orders and receiving none. Instead, he believed that the chromatin is a semifluid colloidal material that acts as part of a connected system and remains continuous with the rest of the cellular material through a network of tiny fibers. The chromatin, and the chromosomes and other sets of parts into which it is organized, therefore act in an important developmental way in the individual in question. Since each individual has two sets of each chromosome, one from each parent, the individual experiences an interaction between the inherited influences of the two parents, in a way such that no two results even of the same two parents will be exactly the same. Chromosomal interactions therefore direct development, yet they also connect the existing individual with the past, through heredity. For McClung, heredity thereby brings continuity but also diversity.

McClung asked the same question that Conklin had raised: how can we explain the existence of different cells given the sameness of chromosomal material? Yet he did not conclude that the cytoplasm also exercises an effect on heredity, as Conklin had. Rather, according to McClung, the cytoplasm is controlled by the nucleus so tightly that "the character of the reaction depends upon the nuclear composition" and, further, that the "nucleus is indispensable in the functioning of the cell."[48]

"This does not at all constitute a denial of other elements of the problem, for which an open-minded attitude should always be entertained," he urged. Yet at least for the time being adopting the chromosomal theory "is a practical measure required by our own mental limitations. To deny or to minimize the value of a consistent body of evidence merely because it is not complete in all details is illogical and unfair. . . . Until some other theory is developed, more consistent with known facts and fuller in its reach, this theory will stand as our best working hypothesis in a most difficult field." For, the theory "stands as one of the highest achievements in biology and offers the most promising guide to further advances."[49] Yet for the moment, McClung offered little in the way of explanation about how chromosomes might effect their influence; he was neither chemist nor physiologist but a morphologist, primarily concerned with chromosomal structures.

In the next essay, Wilson's friend and colleague at Columbia, Thomas Hunt Morgan, took up the related subject of Mendelian heredity. By this time, Morgan had accepted the value of Mendelian genetics, theoretical as it necessarily was. In earlier decades, he had criticized Mendel and Weismann for their unscientific reliance on the existence of hypothetical inherited units, eventually called genes. In fact, Morgan had rather vehemently and insistently rejected the sort of fanciful appeal to invisible germs and to unscientific speculation that Weismann exhibited.[50] He had instead urged the study of living organisms as a whole, with a focus on developing differentiated structures and functions rather than on underlying hereditary factors.

But in the aftermath of his successes with the white-eye male *Drosophila* in 1910, he had decided that "genetics has proved a more refined instrument in analyzing the constitution of the germinal material than direct observation of the germ cells themselves, and while this advance may appear more theoretical than the conclusions based on observations of the cell, this need not mean that it is less reliable." In fact, despite the fact that he had initially been one of the major objectors, Morgan regretted that "the disrepute, into which Weismann's speculations [about quantitative nuclear division] then fell, carried over . . . for a time at least, and prejudiced needlessly the Mendelian situation."[51]

Mendelism involved the law of segregation of factors, law of independent assortment of factors for different characters, and the fact that germ material must consist of discrete units rather than inextricable wholes. Yet since many factors work together to produce a character, the coordinated interactions of parts within the whole remains fundamental. In fact, the genes are coordinated along chromosomes. They do retain some degree of individuality, Morgan believed, but they also interact in some way and to some extent. Morgan regretted that researchers did not at all agree on the way and the extent to which interchange occurs.

After reviewing the Mendelian and chromosomal theories, Morgan suggested that some of the best evidence about chromosomes comes from mapping. This mapping is very like that on a railroad where the reader of the timetable knows the times at which the train is to arrive at a sequence of different stations. By making various simplifying assumptions, the prospective rider can judge the relative distances between the stops. In addition, "Knowledge of the speed of the train and of the condition of the road-bed and of the grades would make it possible to judge more accurately the number of miles between the stations from the number of minutes between the stations."[52] Like the railway passenger, the geneticist must make assumptions but may learn a great deal from carefully considered indirect observations. It is not necessary to have actually taken the ride and directly observed the stations.

As to that recurring question about the relative importance of nuclear and cytoplasmic contributions to development, Morgan answered that it really did not matter and that we do not know enough to give a final answer. Claiming that the cytoplasm is really more important than the nucleus is "an example of obscurantism rather than of profundity." For, in fact, "all the examples of heredity that have been sufficiently worked out show that all adult characters and most embryonic ones . . . are accounted for by the known behavior of the chromosome." In other words, the chromosomes direct whatever happens during development, no matter what the character of the cytoplasm. As a result, "it is clear that whatever the cytoplasm contributes to development is almost entirely under the influence of the genes carried by the chromosomes, and therefore may in a sense be said to be indifferent."[53]

Nevertheless, the chromosomes could not direct if there were nothing to direct. So the cytoplasm is absolutely vital to the actual carrying out of the developmental processes. It might very well be that the cytoplasm is inherited in its own right and divides and differentiates following its own internal inheritance. But for Morgan that remained an open question, since at the moment he knew only that

> the cytoplasm of the eggs of two mutants *may* be as different as are the genes that constitute the chromosome complex of the two mutants; but the cytoplasm in the two mutants *may*, so far as we know, be identical in so far as it changes in reference to whatever kind of genes are present when it develops. On the other hand, the cytoplasms of two types may be different in the sense that in some respects they are affected differently—if affected at all—by the genetic chromosome groups. These questions must be kept entirely free from predilections until we have found out more about the physiological processes that take place in the chromosomes and in the cytoplasm. Whatever the future has in store for us in these respects, the answer does not prejudice the present situation so far as the observed effects of the genes in heredity are concerned.[54]

Although Morgan remained agnostic on the role of the cytoplasm in heredity and development, he did stress the value of pursuing a productive research program in a Mendelian-chromosomal interpretation of heredity.

That Wilson had high respect for Morgan and for the research by his group at Columbia is evident in the contents of Wilson's book. The bulk of new material in Wilson's third edition centered on heredity: on the discovery of accessory sex chromosomes, on which Wilson had worked directly, but also on the role of chromosomes in Morgan's favorite subject, *Drosophila*. Evidence of mutations, of crossing-over, and of alterations in the hereditary chromosomal material held promise for future discoveries of how heredity works, Wilson maintained. Mendelian heredity had begun to seem simple and regular and had brought many divergent facts into a coherent explanation. So Wilson concluded his massive volume with a last sentence sounding a note of hope for unraveling that critical problem of explaining organization. After all, "if Mendelian heredity, at first sight so inscrutable, is effected by so simple a mechanism, we may hope to find equally simple explanations for many other puzzles of the cell that lie beyond our present ken."[55]

Conclusion

Cowdry's volume, wrote one reviewer, is "the largest and most comprehensive ever published on the subject of cytology." It would "stand for many years to come as the most authoritative exposition of a branch of zoology which has grown considerably in recent years."[56] The reviewer noted that the work obviously represented more than the work of any single individual, and indeed that no one person could have covered the breadth of subject matter that the study encompassed. The reviews of *General Cytology* nearly all mention the "exhaustiveness," "extensiveness," and "comprehensiveness" of the work and include other such adjectives stressing scope.

When Wilson's third edition appeared the next year, it also generated rave reviews. Yet, as Conklin admitted in reviewing Wilson's opus,

> probably no single book can ever again deal so comprehensively and judicially with the whole field of cytology. Few other workers are left who were in at the birth of this science and who can speak of its development with the knowledge which comes from intimate contact with persons and problems. It is a monumental work, one of the most complete and perfect that American science has produced in any field, and biologists throughout the world will unite in extending thanks and congratulations to its author on the successful completion of a great work which will always stand as a golden milestone on the highway of biological progress.[57]

Only the unique and aging Wilson could have succeeded, and only then in revising a book rather than in starting from "scratch." No longer could anyone even hope to discuss all of cytology in one book. Cytology—along with biology more generally—had undergone expansion that had carried it beyond the reach of any one scientist or any one approach. Even the leading textbooks reflected widely different methods, interpretations, and emphases.

Wilson's volume offers a compendium of facts, theories, bibliography, and references to an almost incredible range of works by numerous Americans and Europeans concerned with questions about the cell. Yet the focus remains on cell morphology and the central theme is organization: organization of individual cells, of all parts, and of whole multicellular organisms. The volume provided a sustained interpretation of what the cell is, how it arises, and how it works.

Cowdry's volume is much different. It begins with chapters on the chemistry, permeability, reactivity, and therefore general physiological activity of cells; then moves on to questions of structure of protoplasm and cell parts; on to behavior of cells in tissue culture, fertilization, and differentiation; and concludes with chapters on chromosomal and Mendelian heredity. The ordering presents an organized approach to cytological questions. But the volume does not provide the sort of coherent work that Wilson's does. No one praised Cowdry's volume as "monumental." It was, instead, "comprehensive."

The collaborative approach allowed each contributor to focus on his or her strengths and to specialize, while also providing a wide range of subjects covered. The multiple authorship also provided a ground for disagreements or variations of opinion. Rather than presenting a perfectly united front, as if everyone agreed on all points, the collection of essays showed that numerous questions, approaches, methods, theories, and data were acceptable. This was a textbook that showed the interactions and the exchanges within science rather than just a standardized, static result. It left considerable room for suggestions, and for revisions in accepted work.

The collaboration showed that cells could be thought of as little electrical machines, or perhaps as colloidal substances, as sophisticated self-regulating individuals, or possibly as inherited units. Depending on which view one adopted, different things would count as evidence for what was thought to be interesting about cellular structure and function. Different sorts of results would count as knowledge and therefore as legitimate scientific products.

Cowdry's book represented a cooperative effort by researchers who believed that expansion of cytology called for a group of experts to study the whole field. This was, more generally, a time when scientists had begun to move on occasion to the edited volume as a way of presenting a fuller consideration of a number of biological topics. The single-authored textbook remained, of course, but such a text drew heavily on the work of a wide range of other researchers and served largely to systematize the bulk of available information. Such a textbook sought to provide an overview, presenting a package as if to say this-is-what-is-known. The group-generated textbook coverage of a field could provide more depth and more consideration of work in progress and of the best available (but admittedly working) hypotheses; it could also advocate several lines of research from different perspectives rather than being constrained to provide a balanced overview of several alternative viewpoints or an argument for only one. The coordinated conference, the collaborative symposium, and the edited volume increasingly began to come into their own as the century progressed.

Increasing expansion of research into new specialty areas was occurring at the same time and also demanded changes. As more people entered scientific research, and as techniques and problems became more specialized, there were fewer "grand old men" who knew it all. There was simply too much to know, and too many different ways to know it. Groups of biologists had much less in common than they had around 1900, when they would have read a set of the same books and kept up with articles in the same journals. What had briefly come together as "biology" had begun to expand again in new directions.[58] Different problems, different methods, different types of research settings and approaches, and different ways of presenting one's ideas and results further separated biologists into myriad different research directions, as did different audiences and different sources of funding.

As the research enterprise became increasingly specialized and compartmentalized, with people talking less and less to those outside their own special domains, and with fewer people capable of dealing with an entire subject area individually, the collective approach offered a corrective. Properly conceived and executed, collaboration among individual researchers with their different problems, approaches, methods, and presentations could yield products greater than, or at least different from, the sum of the individual parts. Fortunately, to this end, a growing number of research centers also existed which could support such cooperative cross-disciplinary and cross-institutional work. The MBL was one such institution, among many.

These institutions provided a place for scientists to talk to a variety of other specialists with whom they might normally never have discussed their work "back home." Whether summer havens, special research laboratories, or centers where visitors could work temporarily for various periods, places that allowed discussion across the increasingly hardening disciplinary boundaries played important roles in encouraging coordinated and cooperative work. The products reflect the special institutional setting out of which they grew as much as the individuals who contributed. The individuals alone could not have produced the same results, since no one scientist could possibly have been fully familiar with all the varied research represented or have held such a wide diversity of sometimes contradictory views.

General Cytology falls into this category. It could not have been written by one author. Indeed, no one person would have agreed with all of it. Nor would any one person have thought of using all those different approaches or those widely different ways of presenting the ideas. The ways in which Cowdry's book differs from Wilson's illuminates more general changes in biology by the 1920s.

With its variety of perspectives and its substantial guide to the literature in the form of lengthy bibliographies, the volume provided the latest word—or rather a collection of latest words—on cytology but also made it clear that these were just the latest words in a long line of continually revised words about the subject. The effect was to earn the volume considerable attention for several years. Then, new work and new interpretations replaced the old, and the contributors each went on with their own revised contributions. *General Cytology* was not continually reprinted and reissued as a classic. Unlike Wilson's volume, it did not concentrate on the presentation of a valuable array of data and thus serve as a standard reference text for decades to come. Rather, it reflected the best work and an intriguing mix of ideas, methods, and questions for a particular slice of time. It showed the exciting way that cytology had expanded and continued to expand. It illustrated the processes and suggestions rather than the apparently fixed products of science. Whereas Wilson's work was "monumental," Cowdry's was "extensive" and "suggestive." Both reflected the remarkable expansion of cytology in the United States.

Acknowledgments

Thanks to Richard Creath and the participants in this project for their careful readings of earlier drafts and for the many valuable suggestions. Special thanks to Ronald Rainger for generously giving me a copy of Wilson's book when I most needed it.

Notes

1. Leonard Doncaster, *An Introduction to the Study of Cytology,* 2nd ed. (Cambridge: Cambridge University Press, 1924; 1st ed. 1920), p. v.

2. Ibid., p. 265.

3. Edmund V. Cowdry, ed., *General Cytology. A Textbook of Cellular Structure and Function for Students of Biology and Medicine* (Chicago: University of Chicago Press, 1924), Preface, p. v. It would be interesting to know such things as whose idea the volume was, who decided what subjects and authors to include and according to what set of criteria, and other details that might illuminate the nature of and reasons for the project. Unfortunately, correspondence relating to the project at the University of Chicago Press has been placed in the Archives, where it remains closed.

4. Edmund Beecher Wilson, *The Cell in Development and Heredity* (New York: Macmillan, 1925; 1st ed. 1896 and 2nd ed. 1900 as *The Cell in Development and Inheritance*).

5. Wilson, Introduction to Cowdry, *General Cytology,* pp. 10–11.

6. Wilson, *The Cell,* p. xii, explained that he had had no time to respond to Cowdry's volume or to include many references to it since his volume was nearly complete when Cowdry's appeared.

7. Ibid., p. xi.

8. Statistics come from Cattell's *American Men of Science* (various editions) and from other standard biographical notes on each of the individuals.

9. Wilson, *The Cell,* pp. 635, 670–672, 59.

10. Albert P. Mathews, "Some General Aspects of the Chemistry of Cells," in Cowdry, *General Cytology,* pp. 13–95, on p. 15.

11. Ibid., p. 18.

12. Ibid., pp. 21, 22–23, 27, 32.

13. Sir Oliver Lodge, *The Ether of Space* (New York: Harper Brothers, 1909).

14. Mathews, "Chemistry of Cells," p. 90.

15. J. Brontë Gatenby, review in *Nature*, 1925, *115*: 185–187, on p. 186.

16. Philip Pauly discusses a similar point in "General Physiology and the Discipline of Physiology, 1890–1935," in Gerald L. Geison, ed., *Physiology in the American Context, 1850–1940* (Bethesda: American Physiological Society, 1987), pp. 195–207.

17. Robert E. Kohler, *From Medical Chemistry to Biochemistry. The Making of a Biomedical Discipline* (Cambridge: Cambridge University Press, 1982), p. 301.

18. Wilson, *The Cell*, pp. 54–57.

19. Merkel H. Jacobs, "Permeability of the Cell to Diffusing Substances," in Cowdry, *General Cytology*, pp. 97–164, on p. 99.

20. Ibid., p. 155.

21. Ralph S. Lillie, "Reactivity of the Cell," in Cowdry, *General Cytology*, pp. 165–233. Wilson on reactivity, *The Cell*, pp. 101–106.

22. R. Lillie, "Reactivity of the Cell," p. 170.

23. Wilson, *The Cell*, pp. 410–411, 475.

24. Ibid., p. 78.

25. Robert Chambers, "The Physical Structure of Protoplasm as Determined by Micro-Dissection and Injection," in Cowdry, *General Cytology*, pp. 237–309, on p. 238.

26. Edmund V. Cowdry, "Cytological Constituents—Mitochondria, Golgi Apparatus, and Chromidial Substance," in Cowdry, pp. 313–382, on p. 355.

27. Ibid., p. 332.

28. Wilson, *The Cell*, pp. 706, 707.

29. Ibid., pp. 714, 716.

30. Gatenby, review in *Nature*, p. 186.

31. Warren H. Lewis and Margaret R. Lewis, "Behavior of Cells in Tissue Cultures," in Cowdry, *General Cytology*, pp. 383–447.

32. Ibid., p. 429.

33. Wilson, *The Cell*, pp. 102–103.

34. Frank R. Lillie and E. E. Just, "Fertilization," in Cowdry, *General Cytology*, pp. 449–536. Kenneth R. Manning, *Black Apollo of Science. The Life of Ernest Everett Just* (New York: Oxford University Press, 1983), says on p. 149 that "contrary to general opinion Just had done most of the work." And on p. 310, Manning asserts that "Just wrote an equal share, if not more, of the article." Lillie evidently suggested as much in a letter to a potential supporter of Just's research, Julius Rosenwald.

35. Philip Pauly, *Controlling Life. Jacques Loeb and the Engineering Ideal in Biology* (New York: Oxford University Press, 1987), pp. 153–160. Manning, *Black Apollo of Science*, pp. 78–84.

36. Jacques Loeb, "On the Nature of the Process of Fertilization," *Marine Biological Laboratory. Biological Lectures*, 1900, *1899*: 273–282.

37. Pauly, *Controlling Life*, p. 155.

38. Jacques Loeb, *The Organism as a Whole* (New York: Putnam's, 1916), chaps. 4 and 5.

39. Lillie and Just, "Fertilization," p. 522.

40. Wilson, *The Cell*, p. 486.

41. Edwin G. Conklin, "Cellular Differentiation," in Cowdry, *General Cytology,* pp. 537–607, on p. 539.

42. Ibid., p. 599.

43. Edwin Grant Conklin, "The Organization and Cell-Lineage of the Ascidian Egg," *Journal of the Academy of Natural Sciences, Philadelphia,* 1905, *13*: 1–119. He also discussed the theoretical points in "Mosaic Development in Ascidian Eggs," *Journal of Experimental Zoology,* 1905, *2*: 145–223.

44. Conklin, "Cellular Differentiation," pp. 600, 601.

45. Wilson, *The Cell,* p. 106.

46. Ibid., pp. 1085, 1118.

47. Clarence E. McClung, "The Accessory Chromosome—Sex Determinant?" *Biological Bulletin,* 1902, *3*: 43–84.

48. Clarence E. McClung, "The Chromosomal Theory of Heredity," in Cowdry, *General Cytology,* pp. 609–689, on pp. 666, 665.

49. Ibid., pp. 668–669, 682.

50. Thomas Hunt Morgan, "Regeneration: Old and New Interpretations," *Marine Biological Laboratory. Biological Lectures,* 1900, *1899*: pp. 185–208, on p. 191.

51. Thomas Hunt Morgan, "Mendelian Heredity in Relation to Cytology," in Cowdry, *General Cytology,* pp. 691–734, on pp. 693, 714. For more on Morgan and his *Drosophila* research, see Garland E. Allen, *Thomas Hunt Morgan* (Princeton: Princeton University Press, 1978), chap. 5.

52. Ibid., p. 713.

53. Ibid., pp. 727, 728.

54. Ibid., p. 728.

55. Wilson, *The Cell,* p. 1118.

56. Gatenby, review in *Nature,* p. 185.

57. E. G. Conklin, review of Wilson's *The Cell, Science,* 1925, *62*: 52–54.

58. Ronald Rainger, Keith R. Benson, and Jane Maienschein, eds., *The American Development of Biology* (Philadelphia: University of Pennsylvania Press, 1988) explores the establishment of a biological core. Another interpretation appears in Joseph Caron, "'Biology' in the Life Science: A Historiographical Contribution," *Hist. Sci.,* 1988, *26*: 223–268.

Marilyn Bailey Ogilvie

3

The "New Look" Women and the Expansion of American Zoology: Nettie Maria Stevens (1861–1912) and Alice Middleton Boring (1883–1955)

During the first decade of the twentieth century, with professionalization largely accomplished and basic social structures established, zoologists were free to define or redefine their disciplines, explore relationships with sociology and psychology, and add new data through exploration of unknown territories and development of new technologies. Although the potential seemed unlimited, logistical problems sometimes appeared daunting. Structures were built only to become obsolete as zoologists struggled to define their own disciplines, agree on disciplinary interrelationships, and arbitrate "turf" disputes. Social issues, including the place of women in zoology and the export of American biology beyond its continental boundaries, provided additional challenges. These dimensions were as important for defining twentieth-century American zoology as were ideas in rapidly proliferating fields such as behavioral biology, genetics, cytology, systematics, embryology, and paleontology.

This essay addresses some of the constraints and challenges that confronted American women zoologists in search of a career, with a special emphasis on vocational possibilities for these women as American zoology spread beyond North American boundaries. The two zoologists selected as case studies, Nettie Maria Stevens and Alice Middleton Boring, began their careers as students of Thomas Hunt Morgan (1866–1945). Stevens, who published forty papers in cytogenetics, produced descriptive papers that supplied accurate, reliable data to the new field. Her interpretative work on the significance of the chromosomes in determining sex constituted a major theoretical contribution. Boring's forty-eight publications spanned two different fields, cytogenetics and taxonomy (herpetology). An outstanding teacher, she taught numerous Chinese scientists and physicians. Although Stevens and Boring had similar backgrounds, their careers diverged. Stevens, who died young, pursued her career at Bryn Mawr College, making her contributions in the time-honored setting of a woman's college. Boring had a two-part career, first as a teacher with research interests in cytogenetics and embryology at women's colleges and at the University of Maine, and then as a participant in a Rockefeller Foundation–funded program to teach and do research in China, where she changed her field to herpetology.[1]

The Background

Stevens received her Ph.D. degree in 1903, and Boring in 1910. Both were part of a new generation of women scientists who were exploring career options. Professional opportunities, although still limited, were more available in the first two decades of the twentieth century than they had been during the nineteenth century. Women were now able to compete for these positions because some restrictions on graduate education had been lifted. The psychologist Christina Ladd-Franklin worked diligently to ensure that tenacious American women could achieve educational parity in the early twentieth century. Having been refused admission to graduate school at Johns Hopkins in 1878, Ladd-Franklin launched a campaign that proved successful in opening graduate schools to women. However, after they received their Ph.D.s, women were still confronted by the problem of limited employment opportunities. Although women's colleges were the best places for them to find satisfactory employment, the number of professionals these institutions could absorb was limited. As increasing numbers of women received their doctorates, alternative job opportunities were needed. State universities increasingly hired more women but gave them little hope of becoming full professors. Statistics indicated that most women at coeducational schools were instructors, with negligible chances for promotion.[2] The experience of women in government and industry paralleled that of women in academia. In government jobs in the 1920s and 1930s, women were clumped at lower levels, "grossly underpaid," and promoted only rarely.[3] Following World War I, women began to penetrate industry. However, "aware of their limited welcome, these women at first advocated a strategy of stringent self-discipline and cheerful overqualification." The results mirrored those in academia and government—low-paying, low-level jobs.[4]

Women clearly were underrepresented as scientific professionals, but the situation cannot be explained solely by gender bias or marginal competence. Part of the answer involves the change in science from a weakly institutionalized, largely domestic enterprise at the beginning of the nineteenth century to the highly institutionalized, large-scale professionalized activity of the twentieth. Although this trend affected both men and women, evolution toward Big Science had a greater impact on women. A nineteenth-century revision of the role of the family from an economic, financial, and dynastic unit to a nurturing, protective cocoon for its members, encouraged the identification of men with the objective, outside world and women with the subjective, private realm of the home. These changes had an obvious impact on the role of women in science. Increasing professionalization decreed that science previously done by both sexes in the home was no longer quite respectable. Yet because of the new concept of the family, women's roles remained restricted to domestic science.[5]

Membership requirements for scientific societies reflect the effect of the new professionalization on women. Although doors to membership in most scientific societies were formally open by the turn of the century, as Margaret Rossiter notes, ostensible "upgrading" of society standards by creating alternative categories such as "associate" or "junior" members led to fewer women at higher ranks. The practice had an early precedent when the American Association for the Advancement of Science, striving for recognition as a professional society, adopted the practice of two levels of membership. Although the supposed purpose was to keep amateurs of either sex subordinate to professionals in an effort to produce a more prestigious institution, often men who lacked higher degrees were allowed membership in the higher category, whereas women with similar qualifications were invariably placed in the lower.[6]

Other societies followed this pattern. The Geological Society of America considered all its charter members to be fellows, but subsequent fellows had to be voted on by nine-tenths of the ballots cast by all male member's and council. By 1922, of 684 fellows elected, only six (less than 1 percent) were women. This percentage did not represent the percentage of women available for fellowship, for in the 1921 *American Men of Science* women accounted for 5.1 percent of those listed.[7]

The American Ornithologists Union's female members also fell victim to the policy of not electing women to the higher status of fellow. Florence Merriam, in spite of her significant contributions to the field, was not elected a fellow until 1929, seventeen years after she was first nominated in 1912. Margaret Morse Nice made significant contributions to both ornithology and animal behavior. Her research was recognized by the larger scientific community, and she was elected to membership in the AOU in 1931. Yet she was excluded from the "fellow" category until 1937, and her status in various other societies took several interesting turns. She became the first woman president of the Wilson Club in 1937 but was excluded from the all-male Wheaton Ornithological Club. In 1942 Nice was nominated for the editorship of the *Auk* and had considerable support, but the final decision stated that "Mrs. Nice would make a good Associate Editor [but] we can scarcely pick a woman editor for the Auk.[8]

The American Society of Zoologists accepted women members from its origin as the American Morphological Society. The Society's first women members, Cornelia Clapp, Julia Platt, and Harriet Randolph, were elected in 1892 and attended the third annual meeting of the society. However, the number of women members remained small and the dearth of women in power in the society was even more telling. From 1917, when accurate membership lists were established, until the end of World War II, approximately 10 percent of the membership was female.[9] Stevens and Boring had connections with this society, but their organizational opportunities were limited, and even their zoological work was underrated.

Nettie Maria Stevens

Nettie Stevens's family and educational background, professional contacts, and her own initiative combined to allow her to take advantage of the new opportunities open to a woman scientist. But societal constraints limited the extent of her progress. Stevens, the daughter of New Englanders Ephraim Stevens and Julia Adams Stevens (who died when Nettie was two-years old), had a stable background. The family, although not wealthy, had accumulated enough property to guarantee freedom from poverty, and Nettie's stepmother, Ellen Thompson Stevens, proved an appropriate mother substitute.[10] Stevens first attended public schools in Westford, Massachusetts, displaying "quite early an exceptional ability in her studies," and then went to Westford Academy, "where she displayed the same clear visioned aptitude."[11] At Westford Academy, she may have been encouraged to realize that girls could achieve academically, for the institution's philosophy included the statement that "the school should be free to any nationality, age, or sex."[12] Stevens was one of three women to graduate from the college preparatory program between 1872 and 1873.[13] While at Westford, Stevens may have encountered the name Charles Otis Whitman; although Whitman had resigned from Westford before Stevens arrived, he had left a legacy at the school. Their paths may have crossed more directly later in Stevens's career at the Marine Biological Laboratory, where Whitman was director when Stevens worked there during the summer of 1901.[14]

After graduating from Westford, Stevens taught Latin, English, mathematics, physiology, and zoology at the high school in Lebanon, New Hampshire, for three terms. Perhaps it was her savings from this position that made her able to attend the Westfield Normal School (Westfield, Massachusetts), where, according to one account, she passed a four-year course in two years. At this institution she obtained a strong background in the sciences, taking courses in physics, astronomy, chemistry, physiology, botany, zoology, mineralogy, geology, and geography. Her science professors included Joseph Scott, president of the Normal School and a protégé of Louis Agassiz; Frederick Staebner, a graduate of the Columbia School of Mines; and Walter Barrows, who influenced her to continue studying science.[15]

After graduating from Westfield in 1883, Stevens taught school and worked as a librarian before deciding to continue her education at Leland Stanford University. Stanford's reputation as a youthful, innovative school with opportunities for individuals, women included, to pursue their own scholastic interests helped lure Stevens to California in 1896. Given these opportunities, Stevens would have expected Stanford to be a congenial place to work. It no doubt helped when her father and sister followed her there.

During her first year, Stevens proposed to major in physiology, working with Dr. Oliver Peebles Jenkins. However, the next year she switched to histology and started working with Frank Mace MacFarland. During the summer, she pursued her histological and cytological research at Hopkins Seaside Laboratory. She received her bachelor's degree in 1899 and her master's degree in 1900, writing her thesis on "Studies on Ciliate Infusoria."[16]

Although Stevens's sound background prepared her for advanced study at Bryn Mawr College, she did not explicitly state her reasons for returning east. Certainly there were no barriers to obtaining another degree at Stanford, and although it had not firmly established itself as a first-rate institution in her field, neither had the fifteen-year-old Bryn Mawr. Stevens may have realized the unique opportunity that Bryn Mawr offered for the student of cytology or histology. Both Edmund Beecher Wilson and Thomas Hunt Morgan had been employed by this institution. Wilson had left before Stevens arrived, but his ideas and reputation remained. Morgan had already established his reputation as a biologist of note from his work on regeneration and embryology.

After six months at Bryn Mawr, Stevens was awarded a fellowship to study at the Naples Zoological Station and at the Zoological Institute of the University of Würzburg. Such an opportunity had only recently become available to women. Earlier, male zoologists were able to absorb the latest in scientific ideas from Europe, particularly Germany, by participating in research at the Naples Zoological Station. Many institutions, such as Harvard, Johns Hopkins, and the Smithsonian, supported "tables" (places for scholars to work). However, no such resources were available for women until Ida Hyde formed a committee in 1897 to work toward subsidizing a woman's table. In 1898 Margaret Carey Thomas of Bryn Mawr, enthusiastic about the project, headed a committee called the Naples Table Association for Promoting Laboratory Research by Women. Thomas persuaded women's colleges and individuals to contribute fifty dollars to underwrite this research experience.[17]

Stevens took advantage of this new opportunity. During her first visit to Würzburg, Theodor Boveri (1862–1915) was completing his work on the role of the chromosomes in heredity. Studying the development of sea urchin eggs fertilized by two spermatozoa, he concluded that it was "not a definite number, but a definite composition of chromosomes [that] is essential for normal development; . . . this means nothing else than the individual chromosomes possess different qualities."[18] Although no documentation has been found to determine whether this research influenced Stevens's later work, it is probable that it contributed to the development of her ideas. Drawing on the European research experience, she returned to the United States to work on her Ph.D. at Bryn Mawr.[19]

Much of Stevens's early work was descriptive. In her study of the parasitic ciliates, *Licnophora* and *Boveria,* she discovered two new species and compiled details of their life cycles. In addition to her work in taxonomy and on life histories, she experimented in regeneration, following its course in different species, including *Planaria lugubris, Tubularia mesembryantheum, Antennularis ramosa, Stentor coeryleus,* and *P. simplicissima.*[20]

By her selection of appropriate facts and experiments, she was obviously aware of theoretical implications of her descriptive work. While working with *Boveria,* she described four distinct chromosomes in its micronucleus and postulated that the position of the micronuclei indicated their influence over division of the macronucleus. In that study, she isolated many processes that occurred during conjugation and performed experiments to determine patterns of ciliate regeneration. She expanded her interest in regeneration to regeneration in other forms, particularly hydroids and planarians.[21]

Stevens's most important theoretical contribution to science was the chromosomal determination of sex. In 1903 there were two schools of thought regarding sex determination. The proponents of one view argued that external factors were primarily important in determining sex, whereas a second group suggested that sex was determined at the time of fertilization.[22] Stevens's work in 1905 helped establish that chromosomes were responsible for the inheritance of a specific characteristic, sex.[23]

Stevens's role in unraveling the solution to the problem tended to be deemphasized by others. Edmund Beecher Wilson independently arrived at a similar explanation but was more cautious than Stevens in proclaiming his conclusion. Nevertheless, Wilson's role, rather than Stevens's, is often stressed because of his more substantial contributions to the history of biology.[24]

Morgan recognized her achievements, declaring that she "had a share in a discovery of importance and her name will be remembered for this, when the minutiae of detailed investigations that she carried out have become incorporated in the general body of the subject." Yet in the same evaluation of her work, he emphasized her ability as a data gatherer and minimized her theoretical accomplishments.

> Miss Stevens's work is characterized by its precision, and by a caution that seldom ventures far from the immediate observation. Her contributions are models of brevity—a brevity amounting at times almost to meagerness. Empirically productive, philosophically she was careful to a degree that makes her work appear at times wanting in that sort of inspiration that utilizes the plain fact of discovery for wider vision. She was a trained expert in the modern sense—in the sense in which biology has ceased to be a playground for the amateur and a plaything for the mystic. Her single-mindedness and devotion, combined with keen powers of observation; her thoughfulness [sic] and patience, united to a well-balanced judgment, accounts in part, for her remarkable accomplishment.[25]

The fact that Stevens made a significant theoretical contribution to cytogenetic theory and that her contribution was not as generally acknowledged as Wilson's illustrates both the improved possibilities for a woman scientist in the early twentieth century as well as the incomplete nature of her success. Morgan's extravagant praise of Stevens's meticulous experimental techniques and careful observations and his cautious damning of her creativity illustrate the situation. Stevens's career choices and the circumstances of her life both contributed to her success as a scientist and impeded a full realization of her potential.

Despite the significance of Stevens's empirical and theoretical work, she experienced serious career crises. With no support, she was responsible for her own finances. The Carnegie Institution of Washington provided a potential solution to her problem. If she could obtain a grant, she would be able to continue her research. If not, she would be obliged to accept a job that although it would ensure basic support, would not allow her to continue research. In a letter to the Carnegie Institution in 1903, Stevens explained that she had delayed applying for the fellowship because she had hoped to obtain a more lucrative teaching position. But that possibility seemed unlikely since "college positions for women in Biology this year, seem to be very few." . . .

She hastened to explain that "if there were no money question involved," "she would prefer research to teaching. However, practical considerations could not be ignored for "I am dependent on my own exertions for a living and have used nearly all that I had saved while teaching before I began my college-work several years ago."[26]

In the Carnegie Institution's reply to her application, it was stated that funds were exhausted. She renewed her application in November 1903, garnering an impressive group of recommendations. Colleagues, including Wilson, Cornelia Clapp (zoologist and teacher at Mount Holyoke College), and Joseph W. Warren (professor of physiology) urged her appointment. In his letter of recommendation, Morgan emphasized that "Miss Stevens has not only the training but she has also the natural talent that I believe much rarer to find. She has an independent and original mind and does thoroughly whatever she undertakes. I fear to say more but it may appear that I am overstating her case."[27]

In addition to the support of her colleagues, Stevens had the wholehearted endorsement of Bryn Mawr's president, M. Carey Thomas, who wrote to President Daniel Coit Gilman of the Carnegie Institution that "Miss Stevens is one of the few women I know who seems to me to possess original power of a high order."[28]

The supportive letters did not hurry the Carnegie Institution. Finally, Morgan received a letter in March 1904, informing him that Stevens had received the assistantship.[29] The actual grant work, on "the histological side of the problems in heredity connected with Mendel's Law," was completed in 1905.[30]

During that spring, Stevens corresponded with C. B. Davenport about doing research at the biological laboratory in Cold Spring Harbor (Long Island). When she made the inquiry, she was still working under the Carnegie grant. It was not until the next year (1906) that she decided to spend the summer there.[31]

Throughout her life, even as a Carnegie research assistant, Stevens remained affiliated with Bryn Mawr and retained the title Research Fellow in Biology (1902–1904). From 1904 to 1905, she was reader in experimental morphology. Although the trustees eventually created a research professorship for her, she died of breast cancer before she could occupy it. In 1910 Stevens seriously contemplated leaving Bryn Mawr. She complained to Davenport that at Bryn Mawr "we have no opportunity . . . to carry on experiments in heredity." After explaining her desire to juxtapose cytology with experimentation, she asked Davenport for a job—one with "a respectable salary." She did not want people at Bryn Mawr to know of her inquiry, pleading with him "not [to] mention the fact that I have asked you this question as I do not wish it known at present that I am even considering a change." Although when she first made the request, Davenport did not have any job for her in 1911, he offered her a vacant position "beginning the first of July or whenever it would suit your convenience." However, since Stevens's situation at Bryn Mawr had improved, she declined the offer, noting that if it had come a year early she probably would have accepted it.[32]

With her chromosomal theory of sex determination, Stevens made a substantial contribution to theoretical genetics. She also contributed a large amount of factual information to the body of scientific knowledge. These contributions were both facilitated and impeded by a new social setting. As a nineteenth-century American woman, she had opportunities to receive a basic education unavailable to women in earlier times and in different parts of the world. Residents of New England could now afford to supplement basic educational needs with less essential intellectual pursuits. It was not unusual for girls to move beyond the basic subjects to a "higher" education in seminaries and academies. The growing number of women's colleges made higher education possible to many gifted women. Stevens had the advantages of this environment. In her work as a school teacher and librarian, she had the opportunity to continue her exposure to books and ideas. She was also able to save enough money to attend Stanford. From that time on, she lived in the company of the most respected and knowledgeable cytologists of her time. Through her intelligence and industry, she favorably impressed Morgan, Wilson, and others so that they assisted her by exchanging information as well as by writing recommendations for grants.

Yet although her achievements were recognized, Stevens did not reach the level of professional success that one might have expected. She remained on a low rung of the academic ladder at Bryn Mawr, and her theoretical achievements are usually valued below those of Wilson.

Alice Middleton Boring

The second woman considered in this essay, Alice Middleton Boring had a background similar to Stevens's, but her career diverged in a rather radical way. After embarking on a traditional career in biology, she became part of a new trend in early twentieth-century biology—the export of science to Asia. Through the largess of philanthropic foundations, particularly Rockefeller, American scientists and physicians began to make careers in China. As Mary Bullock noted, a different spirit possessed those Americans who crossed the Pacific than those who crossed the Atlantic. Whereas travelers to Europe were eager to identify with the culture of the Old World, those who traveled to Asia generally accepted the premise that the Far East was "backward, illiterate, impoverished, and heathen." Although there were many reasons that Westerners went to the Far East, including religious fervor and individual aggrandizement, most accepted the superiority of Western civilization.[33]

Like Stevens, Boring began her career by working in two important areas of early twentieth-century zoology, developmental biology and cytology. Boring was also unmarried and responsible for her own financial support. Working within the system, she obtained positions both at women's colleges and at a state university. When the Rockefeller Foundation made it possible for American scientists to work in China, Boring opted for a career change that shifted her research interests from experimental biology to herpetological taxonomy, thus linking herself with another early twentieth-century trend, classifying the world's unknown organisms. Finally, through her teaching, she helped facilitate the adaptation of assimilation of Western science by the Chinese.

The daughter of a pharmacist and one of four children, Boring attended Friends' Central School in Philadelphia, a coeducational college preparatory school, and graduated in 1900.[34] She continued her education at Bryn Mawr College, where she received all her degrees. Boring first encountered Morgan in her freshman biology class. Although her grades were not outstanding, she admired him greatly, writing that "as a freshman I was enthralled by his telling us humble students of his Minor Biology Class about the exciting things he had been doing with frogs' eggs the night before." During Boring's junior year, Morgan allowed her to participate in his experiments. Flattered by his trust, she claimed that in this year she was "sold to biology for life."[35]

Although Boring remained at Bryn Mawr to work on her master's degree, she may have found her first graduate year anticlimactic, for Morgan had left for Columbia University. Because Bryn Mawr was not as exciting for Boring without Morgan, when the time came to continue her graduate work she decided to go to another institution close to home, the University of Pennsylvania. At Pennsylvania, she studied with another outstanding scholar, Edwin Grant Conklin (1863–1952). He never attained Morgan's renown, but his work in cytology, embryology, and genetics was well respected. Conklin not only was Boring's teacher but was a close friend. Perhaps Boring would have remained at Pennsylvania for her doctorate had Conklin stayed, but he moved to Princeton. As a result, Boring returned to the Ph.D. program at Bryn Mawr in 1906, where Stevens supervised her dissertation.[36]

Boring's work fit into the mainstream of American biology. During the early twentieth century, the convergence between the behavior of chromosomes in cell division and Mendelian experimental breeding data was investigated extensively. Although Stevens's and Wilson's independent investigations yielded evidence that sex in particular insects was determined by chromosomes, additional research was needed before generalizations could safely be made. At Stevens's suggestion, Boring provided supplementary support in her dissertation research, "A Study of the Germ Cells of Several Hemiptera Homoptera," a project she began during her master's work and continued with Conklin at Pennsylvania.[37]

The academic year 1908–1909 Boring spent studying at the University of Würzburg and at the Naples Zoological Station, continuing her morphological and physiological work. Like Stevens and many other American zoologists, she went abroad to bring home new ideas gleaned from European investigators. Also like Stevens, she worked with Theodor Boveri and published two papers based on her European work.[38]

After she returned to the States, Boring began her teaching career at the University of Maine. She seemed well on the way to establishing herself as a university professor with excellent research credentials. She was guided by the interests of Raymond Pearl (1879–1940), then chairman of the Department of Biology at the Maine Agricultural Experiment Station (1907–1916). Although Boring collaborated with Pearl on several publications, she never became involved in his earlier fascination with the power of eugenics to solve social problems. By the time she worked with him, Pearl had begun to doubt the effectiveness of selection for the "improvement" of organisms, including humans. He had mounted an attack on both natural and artificial selection as viable mechanisms for producing new types. Boring apparently did not become involved in the controversy, and her contributions remained basically cytological and histological. Nevertheless, at the agricultural station Boring used Pearl's favored research animals, poultry and cattle, as research subjects.[39]

In her position at Maine, Boring advanced more rapidly than most women. She did not follow the previously noted pattern, where most women in the state university environment remained at the instructor rank. Boring advanced to associate professor after four years, but she stayed at that rank for her remaining five years.

Up to that point, Boring seemed to have embarked upon a typical career pattern for a scientist. However, in 1918, the pattern changed. Whether Boring was dissatisfied with her own performance, the support at Maine, or the reactions of her colleagues is not known. At any rate, she decided to leave Maine to accept a temporary appointment as instructor in the premedical department of Peking Union Medical College (PUMC) in China. Boring would only remain in China for two years in this position, but during this time she would begin to recognize an alternative career path.

Evidently her decision to leave for China did not stem from a belief that her advancement at Maine had been stifled because she was a woman, for her promotion to associate professor had been rapid. Her assessment of her own research potential may have contributed to the decision. Although she had published solid research papers, Boring did not consider herself a creative research scientist. Yet she recognized that she was a good teacher and enjoyed the opportunity to devote time to serious students. The service orientation engrained in her from her Quaker background provided an additional reason for going. Finally, she had a real concern that life was passing her by and that if she was to make a difference, she needed to make a change.[40]

Boring's new experience in China was made possible by the Rockefeller Foundation's decision to support medical education in China. After the foundation received its charter from the state of New York in 1913, individuals with diverse interests (business, religion, teaching, and government) met "to consider whether this new agency might be able to carry out a useful work in China, and to advise as to the kind of activity, if any, it might most advantageously undertake." They concluded that "aid in the development of modern medicine" would be an effective service. The Rockefeller trustees sent a commission to China to study the situation.[41] The 1914 report of this commission acknowledged the need for better medical education. The foundation accepted the report, and a special department, the China Medical Board, was organized to take charge of the work.[42]

When the first China Medical Commission met in 1914, it attempted to determine the ideal nature of medical education in China. The physicians Abraham Flexner and Charles Eliot disagreed about the purpose and programmatic implementation of medical education in China. Flexner contended that medical schools could best serve China's medical needs by providing a broad-based practical medical education. He argued that medical education in China need not be based on a core of the basic sciences, for "an immense amount of medical treatment can be practiced with a very limited knowledge—or perhaps no knowledge in a wide sense of basic science—of chemistry, physics or biology."[43] Eliot, on the other hand, insisted that medicine must be firmly grounded in the basic sciences. His arguments were persuasive, and eventually Flexner conceded that it was both desirable and possible to provide a premedical education equal to the best programs in the United States. The elitist implications of this decision continued to haunt the program through 1949.

After surveying medical education throughout China, commission members concluded that Peking was the best site for the proposed new medical college and that the old Union Medical College (a mission institution) should be purchased and incorporated into the new structure. When PUMC was opened, John D. Rockefeller, Jr., described the facility as "the best that is known to Western civilization not only in medical science but in mental development and spiritual culture."[44] Entrance requirements, curricula, and training objectives comparable to those in the finest medical schools in the United States and Europe were introduced into China's traditional agrarian society. The program was designed to train a small number of scientifically oriented medical doctors, who, in turn, would train other doctors.[45]

In response to the commission's recommendation, the new president of Peking Union Medical College, Franklin McClean, supported the establishment of the premedical school as the first program of the new medical college. On December 20, 1916, this program, which provided Boring's initiation to China, was approved, and the premedical school opened on September 11, 1917. Almost one year later, in July 1918, Boring received notification that she had been appointed "Assistant in Biology in the Pre-Medical School of the Peking Union Medical College for a period of two years beginning August 1, 1918, at an annual salary of $1,200.00. You will also be furnished with a residence in Peking, probably together with some of the other single ladies of the school."[46]

Boring found the China experiment liberating. In the new environment, she had an opportunity to be a part of an institution that was just beginning to define itself, could operate in a setting where academic constraints were not as well defined as they were in the United States, and where women played an important role. In addition, and to Boring more important, she might have a unique opportunity to spread Western scientific knowledge to China, thus helping the Chinese help themselves.

Although not particularly sympathetic toward missionaries, Boring, may have reaped some benefit from the acceptance of women within their ranks. Women clearly dominated the lay cohort of missionaries in China. A typical group of 100 missionaries in 1910 would have included thirty ordained men, twelve laymen involved in nonmedical work, five physicians (including one woman and one of the ordained men), and fifty-five women. Women were clearly subordinate in the hierarchy but had a considerable amount of practical influence. At first, mission boards had hesitated to send out single women, but by Boring's time, the practice had been well accepted for a number of years.[47] Originally, she was the only woman on the faculty (Alice Ryder, assistant in English was appointed late in 1918), although two others, Helen R. Downes in chemistry and Edna M. Wolf in biology, came in 1920, the year Boring left.[48] Later, other women were involved in the university's nursing programs.

The selection of students was also nondiscriminatory in the original program. Franklin C. McLean wrote that "women are to be admitted to the Pre-Medical School on the same terms as men."[49] In spite of this liberal philosophy, university officials clearly were not planning for many women students. One report indicated that since women would be constituting only a small fraction of the entire student body, "it would be expedient for us to make some arrangement for them entirely apart from the provision laid out for the men: as for instance, through the YWCA or a similar organization." The report further indicated that the "fact of having a few women students in attendance" would not justify "buying the costly ground and erecting an entirely independent unit for gymnasium purposes."[50]

The employment of Alice Rohde as associate professor of pharmacy illustrates the varying attitudes toward women professionals found at PUMC. Although Rohde had applied for the pharmacy position, Edmund V. Cowdry considered her unsuitable on the ground "that she had moved so many times during her scientific career . . . in addition to her somewhat mature years." He indicated "that it would be wiser to select one or two young men to carry on the work in a temporary fashion until a more thoroughly suitable candidate for the higher rank could be secured." Roger S. Greene, however, disagreed with Cowdry's assessment. He argued for Rohde's appointment from the standpoint for both expediency and acceptability. The position needed to be filled immediately; pharmacology was a "somewhat restricted" field; and it might be hard to find suitable young men to fill a junior appointment and come to China on short notice. Since the New York officers had reviewed Rohde's qualifications and pronounced her acceptable, Greene was in favor of hiring her. To mollify Cowdry, however, he agreed to have the matter discussed at an informal meeting of the administrative board and then to cable the results of the meeting to New York. Greene's arguments were accepted and the board "RESOLVED that it be the sense of this meeting that a cable should be sent approving the appointment of Dr. Alice Rohde as associate professor of pharmacology."[51]

Although the issue of women did not preoccupy Boring, her research suffered during this time. For the most part she was not concerned, for two years was not long enough to get involved in a major project. It did give her time to reflect on her abilities and interests. As noted, she had produced important work in cytogenetics before leaving for China. Nevertheless, when she examined her output she fretted that, although she had the ability to collect data accurately and to interpret results with understanding, she might lack the spark of originality that would transform her research from the mundane into the creative. Participation in the PUMC experiment gave Boring an opportunity to examine her own strengths and weaknesses.[52]

Boring's first experience with teaching in China was positive. She found her colleagues congenial, making the adjustment more pleasant. Since Charles R. Packard, who became head of the Biology Department, had studied under Morgan at Columbia, she made a friend with whom she could share experiences and interests.

When the two years were over, Boring was reluctant to leave China to begin a new teaching position at Wellesley College. PUMC Resident Director Roger S. Greene reported in his diary that he had had a conversation with Boring in which she reported that "she has not attempted to do any research work while in Peking, as she realized that she was to be here only for a short time and did not feel that there were any special subjects on which it was worth while to begin work." However, she made it clear that she would entertain different alternatives in order to work in China. She told Greene that she had "gained some very valuable teaching experience in dealing with Chinese students . . . and feels that her stay here has been most enjoyable and worth while, and suggested the possibility that if we required a dean of women some time in the future, she might consider an offer."[53]

She did return to teach at Wellesley but was unhappy and remained open to any opportunity to return to China. The chance occurred when she heard of an opening to teach biology in a proposed new institution, Peking (later called Yenching) University ("Yenching" will be used when referring to this institution). The initial steps in the formation of the university took place before Boring arrived in China to teach at the PUMC premedical school. Intense negotiations had occurred between 1916 and 1918 to establish a federated university forged from four separate entities. The largest of these institutions was Huei Wen (Peking University), a school established in 1870 by the Methodist Mission in China. The second institution was a men's college, North China Union College located at Tungchou, thirteen miles east of Peking. As its name indicates, this college was a product of cooperation in higher education. It represented the Congregational Church through its American Board Mission, the Presbyterian Church through its Mission, and the British Congregational Church through its London Mission. The North China Union Woman's College, like its men's counterpart, was a constituent college in the North China Educational Union. It had the earliest origin of the three, developing from a girls' school, the Bridgman Academy, established in 1864. The School of Theology, itself a composite institution, constituted the final ingredient to be merged into the reconstituted university.

Once the principle of union was accepted, the Methodists and the Presbyterian-Congregational group became mired in petty disputes. The spirit of Christian goodwill appeared notably lacking as negotiations toward the union continued. Disagreement over name, location, and almost everything else made the consummation of the union seem highly unlikely. It was not until John Leighton Stuart (1876–1962) agreed to accept the presidency in 1918 that the disagreements between the two institutions ceased. "Immediately," wrote Stuart, "I began to realize how much more intense were the divergencies between the two groups than I had expected."[54]

The opportunity to become part of this enterprise appealed to Boring. China was seldom far from her mind while she was teaching at Wellesley. She kept in touch with her former colleagues and friends and, consequently, was aware of the trials and triumphs involved in the establishment of the new university. When Dean Alice Frame suggested to Boring that "it would be a great help to them if Wellesley would lend me to them for two years to teach biology," she saw a chance to leave Wellesley without burning any bridges.[55] If Wellesley agreed to give her a two-year leave of absence, she would have an opportunity to teach and do research in her beloved China without jeopardizing her security. Making a career in China still had not occurred to her or to those on the China Medical Board who would employ her.

Although she assumed that the position would be temporary, Boring wanted to make absolutely certain that she understood every aspect of the situation before she made a commitment. In order to clarify her understanding, she queried everybody who would listen. Greene, however, bore the brunt of the assault; she pressed him unmercifully for additional information. Eventually she was satisfied that she had the answers to the pivotal questions, including the relationship between the biology departments at Yenching University and PUMC, the acquisition and disposition of equipment, the relationship between the women's department and the rest of the university, the relationship between the faculties of the two institutions, and salaries. It was President Stuart who finally was able to allay her fears.

The issue of Boring's relationship to the existing biology faculty at PUMC was more difficult to resolve. During the period of negotiations, Greene and Stuart both recommended a single administrative chair for the biology departments of PUMC and Yenching. Although PUMC Medical Director Henry S. Houghton proclaimed that Stuart's goal was to secure Nathaniel Gist Gee for the position, both he and Stuart agreed with Greene that Boring was well qualified to administer both departments on an interim basis.[56] In spite of his irritation with Boring's incessant questions, Greene served as her advocate with Houghton. After apologizing to Houghton for offering Boring the position without his approval, Greene stated his opinion that Yenching "could hardly do better than ask Miss Boring to take charge of the teaching of biology for two years," and that she should not be expected to teach under anyone with less experience than she had. Although Greene did not explicitly state that Boring should be the head of the departments at both Yenching University and the medical college, he implied this relationship, for the dean of the Biology Department at PUMC, Aura E. Severinghaus, was younger and far less experienced than Boring. To Greene's relief, Houghton praised Boring as "an unusually capable and stimulating teacher" and expressed his hope that she would accept the position for two or three years. He remained silent about the headship of the combined departments. Stuart, however, made it clear that the biology staff of both institutions would be under her direction.[57]

It came as a shock when, after Boring had officially accepted the position, she received a letter from Anna Lane, Yenching University senior biology teacher, stating that "Mr. Severinghaus is taking over the headship of the Biology Department" for PUMC and Yenching. That "is what I thought I was doing," Boring claimed.[58] Astonished and dismayed that even after her meticulous questioning of every detail something had gone amiss, she sought an explanation. Both Greene and Stuart had indicated that she was to be the combined head. Since Greene and Stuart both favored Boring, it must have been Henry Houghton who balked.

Why, Boring asked Stuart, "would Dr. Houghton have done this when he knew your negotiations with me were pending? Certainly, she did "not want to be petty and spoil everything," . . . but she nevertheless insisted that professionally she could not work "under a man so much younger and less experienced than I am." Hurt by Houghton's rejection, she still recognized the importance of cordial relations between the two institutions and proposed an alternative. She suggested that Severinghaus become the head at PUMC and she at Yenching so that they could "coordinate and cooperate." Although she would have preferred the arrangement that Stuart had preferred because it would place Yenching in a stronger position, she was willing to accept the alternative, for she had been assured that "Severinghouse [sic] is a fine fellow and wants to do everything to help us."[59] Apparently Houghton did not want to consider the possibility of PUMC being subordinate to Yenching in any respect. It is also likely that he would have found it uncomfortable to have a woman in that position. Houghton and others had considered the compromise that Boring proposed and were willing to accept it.[60]

When Boring left for her temporary appointment at Yenching University, her primary concern was to build a first-rate biology department. She understood that organizing the department, supervising the move (the university was still located in its old quarters in the city), preparing for class, and advising students (she soon became the premedical adviser) would leave little time for research; still, she may have recognized that, if she chose, research opportunities in genetics and embryology would be available in the future. The two foundations pivotal in determining the direction of scientific research in China, the Rockefeller Foundation with its China Medical Board and the China Foundation for Education and Culture (the body organized to supervise the disbursement of the American Boxer Indemnity balances), seldom funded research in genetics, but they made exceptions for Yenching and Nanjing universities. They anticipated that this research would lead to improved crops. Yenching was favored by the China Medical Board because of its association with the Rockefeller Foundation's Peking Union Medical College.[61] Boring's correspondence, however, makes it clear that her decision to go to Yenching was not influenced by the possibility of doing research in genetics.

If Boring had been committed to continuing her research when she arrived in China, she would have been thwarted. Even though the potential for research existed, the political situation in China of 1923 made research of any kind difficult. Her reintroduction to China was sandwiched between two major periods of student demonstrations, as well as a civil war between competing warlords. Not surprisingly, the unrest resulted in the dilution of her energy and interest in research. Nor, given her Quaker idealism, was it unexpected to find her identifying more with the struggles of the Chinese people than with research that might or might not have practical implications. By the summer of 1925, Boring acknowledged to herself and to her family that she wanted to make her career in China. She could not announce it officially, because she was not sure that remaining permanently in China was an option.[62]

Everything that Boring had been working toward since returning to

China would be reaching a climax in 1927 when she was scheduled to return home. Because of the delay in the completion of the campus, the Biology Department would still be moving into its new quarters at about the same time she was to sail for North America. Returning to Wellesley when she was needed in Yenching to move equipment from the old laboratory to the new, inventory apparatus, order and catalog materials and supplies, and organize the department seemed inconceivable. Her proprietary interest in the department's operation made her unwilling to leave until certain that it would function smoothly. "After I have planned all those laboratories and storerooms, I want to move in and see things working in them before I leave."[63] She asked for and received an extension of her leave of absence, so that by the time she actually took her furlough (summer of 1928) Stuart had assured her that her position at Yenching was secure. "It is fate," she wrote. "I belong here. I know how to get along with this crowd. I understand their standards and evaluations in life and they understand mine and this is true of both Chinese and foreigners. I shall not become as good a scientist out here as I might have in America, but I am a much better developed human being, and I always was a better teacher and that is what is needed in China in the present stage of her science."[64]

When she returned home, Boring was not eager to encounter her old research associates. Although the turmoil of her first years at Yenching had left little time for research, she had decided to pursue a new and different line of study—to move from experimental biology to taxonomy. Since her former research associates viewed experimental genetics and cytology as exciting biological frontiers, Boring was convinced that they would not approve of the shift to the less glamorous area, taxonomy, that she had already determined to make. For that reason, she failed to visit her old friend and colleague Dr. Pearl, and only informed him of the change in her research focus in August 1929, as she was "steaming across the Pacific back to China after my year in America."[65] Pearl, though he apparently thought that one could not do first-rate experimental science in China, nonetheless maintained that "you [Boring] have done a very wise thing in turning your energies in this direction."[66]

Several factors influenced her research shift. In China she became very involved in teaching and in advising premedical students, and becoming a first-rate experimental scientist involved a more complete dedication to the subject than she cared to muster. Fascinated by the variety of the Chinese herpetofauna, she decided that she could do taxonomic research sporadically as time constraints allowed. In studying Chinese reptiles and amphibians, Boring was engaging in a less competitive form of research than in her previous experimental work, an area in which she found her own achievements wanting. By working on taxonomic problems, Boring could expect that her contributions would transcend the mere exporting of American science.

The probability of discovering information inaccessible to most Western biologists made up for any inadequacies she might have felt about her own research abilities. Still, she was unsure of the response of her experimentalist friends, and although not deliberately avoiding her former colleagues, she did not actively seek their companionship.

At the same time she avoided the experimental biologists, Boring sought the advice and opinions of the herpetologists Clifford H. Pope and Gladwyn Kingsley Noble of the American Museum of Natural History. During her years in China, Boring had met Noble and Pope and supported them in their work on Chinese reptiles and amphibians. She was able to advise them about conditions in the country, provide sources for equipment and supplies, and put them in touch with individuals who could assist them. The friendly collegial relationship that they had established continued when Boring returned to the United States on leave. Instead of returning to Wellesley as she originally had planned, she spent her leave at the University of Pennsylvania. There she corresponded with Pope and Noble while she worked "hard at anatomy and animal distribution" at the University of Pennsylvania.[67] From Pope and Noble she received reprints of papers, explanations of terms, and unlimited access to the facilities of the American Museum.

Her taxonomic work at the American Museum had immediate results. She finished "a complete list of all the Amphibia ever recorded from any part of China,"[68] as well as a bibliography and notes on distribution. In the fall of 1929 she wrote Noble that she and N. Gist Gee were presenting a paper at a biology conference held in conjunction with the formal opening of Yenching University and expected to publish it with the Peking Natural History Society.[69] Noble, pleased with her results was especially complimentary about her success in interesting students in the study of Amphibia.[70]

Boring recognized that the relatively unknown Chinese fauna offered a unique opportunity to explore organic diversity. William J. Haas makes the case that during the 1920s and 1930s, taxonomy probably became the most important scientific discipline to be practiced in China, partly because it offered an opportunity for a reconnaissance mission to a part of the world not well known to Western science. Although Rockefeller adviser N. Gist Gee tried unsuccessfully to induce the foundation to support taxonomic research, it did so indirectly. As Haas observed, it made such research possible by funding physical plants and salaries at selected universities. When the Rockefeller Foundation gave money to biologists who did research in taxonomy, "it did so without regard to these biologists' research interests," but by disbursing the money to them as teachers. They then used the money for research. Taxonomic research was more directly funded by the China Foundation for the Promotion of Education and Culture. It provided direct funding for research and supported institutions studying the flora and fauna of China.[71]

To early twentieth-century taxonomists, Asia was uncharted ground. In the first two decades of the century, explorers garnered financial support, braved physical hardships and political turmoil to proudly present new specimens to their museums.[72] Roy Chapman Andrews of the American Museum led two of these early expeditions. Although satisfied with the limited accomplishments of the first two expeditions, he had a more ambitious program in mind for a third one.

Although science was ostensibly the focus of these expeditions, a cultural mystique lurking beneath the surface directed the interest in collecting. As Donna Haraway has indicated, the fear that decadence was eroding the substance of male, capitalist, imperialist society motivated the three public activities of the American Museum: exhibition, eugenics, and conservation. By collecting and preserving specimens, dwindling nature could be fixed "in the face of extraordinary change in the relations of sex, race, and class."[73] In the very act of collecting, the rightful male role of the domination of nature and of his own sphere could be played out. If after struggling with recalcitrant nature the scientists-explorers could prevail and wring her secrets from her, the sovereignty of man could be confirmed. The same motivation may have inspired the cultural imperialism of the Western countries.

Andrews's ambitious program for the Third Asiatic Expedition fit into the program of confirming the superiority of man. This expedition was especially concerned with a search for evidence of human evolution. By bringing together biologists from many different areas and correlating their expertise, he hoped to solve some puzzling problems. Museum director Henry Fairfield Osborn believed Asia was the birthplace of *Homo sapiens,* so much of the expedition involved a foray into the Gobi Desert of Mongolia.

Less colorful taxonomic activities benefited from the expeditions as well. Although the major thrust of the Third Asiatic Expedition (1920–1930) involved an expedition to Mongolia, it also included a survey of Chinese fish, amphibians, and reptiles. In 1921 Pope, the museum's herpetologist, arrived, at first unable to speak Chinese. Since Mongolian winters, with temperatures dropping to forty or fifty degrees below zero, severely limited the numbers of reptiles and amphibians available for study in this area, Pope did not go to Mongolia. He remained in China to make a survey of herpetology and icthyology in each of its provinces that resulted in the largest collections of fish, reptiles, and amphibians ever made in China.[74]

Boring became interested in helping Pope with his survey. Recognizing that cooperation was the basis of success in taxonomic work, she provided specimens and data. He, in turn, loaned her specimens from the collections of the American Museum. Such an exchange ensured that each could examine types, topotypes, and paratypes and contributed significantly to understanding the geographic distribution of species and subspecies in China.

In the fall of 1925, the Peking Natural History Society was founded with Boring as a charter member. The new organization reflected the burgeoning interest in collecting, collating, and classifying that flourished during the heyday of the Third Asiatic Expedition. The society proposed to foster the systematic study of the flora and fauna of China, and its publication outlets for its members' research made it especially valuable. The society produced two types of publications: a quarterly bulletin of technical studies on fauna and flora and numerous handbooks of a more popular nature.[75] Boring took advantage of both the *Bulletin* and the handbook series to publish the results of her research.

Although the society's publications provided an important outlet for her research, the networking activities that it encouraged were even more important. Without contacts with other collectors, other museums, and other biologists, taxonomic research would have been impossible. The paucity of a long series of specimens from contiguous localities hampered herpetological taxonomy, and therefore Boring needed to examine every specimen available in order to determine specific relationships. As she and her student C. C. Liu found when they began to study *Kaloula,* an Indomalaysian genus of frogs, the records were very confused. Through the cooperation of students in Nanking and Taiku, Shansi, records of collectors Leonard Stejneger, Clifford Pope, Karl Schmidt, and others, Boring and Liu were able to reinterpret Pope's assessment that *Kaloula* from central China represented a different species from *Kaloula* of northern China. By comparing specimens from the Yenching campus, from Taiku in Shansi, and from Nanking, with descriptions of materials from other authors, they concluded that all forms represented a single species. To Boring and Liu, the measurements and color differences did not seem significantly different to warrant calling them different species.[76]

By 1936, Boring was scheduled for another furlough. This time, she was more willing to return to the United States, for she needed to compare types and topotypes at the American Museum with her specimens in China. Only by such a comparison could she positively identify her specimens and postulate specific or subspecific relationships. The China she returned to in 1937 was not congenial to research: China and Japan were in a state of open although undeclared warfare. The events of 1937 changed the nature of both Boring's research and personal life. Although the Yenching University campus was exempt from many of the "petty tyrannies" perpetrated by the occupying force, she reported that a sense of foreboding hung over all her activities. By 1939, it was very difficult either to receive or to send mail, and everyone had financial problems. Boring loaned money to friends until she found she had none left. Plans were made to evacuate foreigners from China, but Boring elected to remain, explaining that "this is home."[77]

For years, Boring and Pope had planned a monumental collaboration on Chinese amphibians to correspond to Pope's *Reptiles of China.* Boring's research would have been central to the project, but world events dictated that the major monograph could not be published. Supporters were unwilling to pour money and supplies into collecting expeditions in a turbulent China. Both Pope and Boring realized that Boring's days in China were numbered, and that haste was important if anything was to be published before the inevitable retreat. Although disappointed, Boring agreed that they should publish their results in an abbreviated form in the *Peking Natural History Society Bulletin*. The resulting publication, which included characteristics, distribution accounts, and keys for each taxon, proved a valuable addition to the taxonomic literature of China.[78]

Clearly the decision to publish an abbreviated account was a wise one, for in December 1941, following its attack on Pearl Harbor, Japan closed Yenching University. Yenching faculty members were interned in Beijing. During this time Boring worked on her bibliography of Chinese Amphibia. She wrote to her friend Grace Boynton reporting that she had received fifteen pages of galley proof and was "busy making maps for this paper." In March 1943 the British and American faculty were moved to a concentration camp in Shandung Province. Boring's brother, sisters, and friends lost contact with her for over a year, but in the autumn of 1943, she was among a group of Yenching University staff sent back to the United States. In a letter written during the voyage, Boring announced with characteristic cheerfulness that "we have been marvelously well. . . . We shall not look like physical wrecks when you see us in New York, even if our clothes may be rather dilapidated."[79]

Boring took a post as instructor in histology with a research assistantship in the College of Physicians and Surgeons, Columbia University (1944–1945), then spent a year as visiting professor in zoology at Mount Holyoke College (1945–1946). She found the research at Columbia tiresome, and teaching young American women at Mount Holyoke unrewarding. She availed herself of the first opportunity to return to China, taking up her duties at Yenching University in the autumn of 1946. Although the university had made "a marvelous comeback," politics again intruded on the educational scene. The conflict between the Communists and the Nationalists grew into a war. Boring eventually accepted and even felt some enthusiasm for the Communists' approach. In 1949 she wrote that she was "surprised to find that in spite of my opposition in the past, I now am full of hope!"[80]

It was not disenchantment with the new regime but an accident suffered by her sister, Lydia, that occasioned Boring's return to the United States in 1950. After her return, she lived close to her brother Edwin and his family in Cambridge, Massachusetts. Although she accepted a part-time professorship of zoology at Smith College in Northampton, Massachusetts from 1951 to 1953, she spent most of her time keeping in contact with her old friends from China, participating in activities connected with China, and volunteering her services for various charitable causes. After suffering several attacks, she died of cerebral arteriosclerosis in 1955.[81]

Conclusion

Nettie Maria Stevens and Alice Middleton Boring were both part of mainstream American zoology. They were professionals in the best sense of the word. They possessed doctorates, studied under well-known teachers, published in professional journals, participated in scientific societies, and earned their livings from their science. Their scientific interests coincided with those of their colleagues. The relationship of the chromosomes to heredity, eugenics, and reproductive biology were all important topics to early twentieth-century zoologists. Stevens and Boring worked on one or more of these topics. Boring added another dimension, as she participated in another important aspect of zoology at this time—expanding the purview of American zoology beyond the North American continent with the support of the Rockefeller Foundation and classifying hitherto undescribed animals of China.

Stevens and Boring did indeed have characteristics in common with all good scientists of the time. Yet because they were women, they encountered special barriers as they considered career options. Although certain positions were clearly unthinkable, other jobs had opened up because of pioneer work done in the nineteenth century by such women as Christina Ladd-Franklin. Both women made fortuitous career decisions. Stevens had little choice. Women's colleges provided the best job opportunities for women scientists. Stevens had only one professional opportunity to leave Bryn Mawr but did not accept it because her situation at Bryn Mawr had improved. Boring tried both the women's college and state university options. When neither was to her liking, she found a third alternative—teaching at Rockefeller-funded institutions in China.

The careers of these two women illustrate both the progress and problems facing twentieth-century women zoologists. The struggle of women to be fully accepted as professionals was clearly a part of twentieth-century zoology. The expansion of zoology from male professionals to male and female professionals occurred during this period. At the same time that zoology began to accept women professionals, it also expanded across the sea with the aid of the foundations, particularly the Rockefeller Foundation—a geographical expansion that added a new career option for women zoologists.

To succeed as a professional, women had to be talented, diligent, and lucky. The opportunities were limited, but they were present. Alice Boring showed a special type of creativity by developing a career for herself in China.

Notes

1. For additional information on Stevens, see Marilyn Bailey Ogilvie and Clifford J. Choquette, "Nettie Maria Stevens (1861–1912): Her Life and Contributions to Cytogenetics," *Proceedings of the American Philosophical Society,* 1981, *125:* 292–311.

2. Margaret W. Rossiter, *Women Scientists in America. Struggles and Strategies to 1940* (Baltimore: Johns Hopkins University Press, 1982), pp. 31, 110.

3. Ibid., p. 218.

4. Ibid., pp. 265–266.

5. Pnina G. Abir-Am and Dorinda Outram, eds., *Uneasy Careers and Intimate Lives, Women in Science, 1789–1979* (New Brunswick: Rutgers University Press, 1987), pp. 3–6.

6. Rossiter, *Women Scientists in America,* p. 275.

7. Ibid.

8. Marianne Gosztonyi Ainley, "Field Work and Family: North American Women Ornithologists, 1900–1950," in Abir-Am and Outram, *Uneasy Careers and Intimate Lives,*" pp. 67–70; Mitman and Burkhardt, this volume.

9. Keith R. Benson and Brother C. Edward Quinn, "The American Society of Zoologists, 1889–1989: A Century of Integrating the Biological Sciences, *American Zoologist,* 1990 *30:* 353–396. On other organizations, see Rossiter, *Women Scientists in America.*

10. Ogilvie and Choquette, "Nettie Maria Stevens," p. 294.

11. Obituary Notice, unknown newspaper source. From Fletcher Library, Westford, Massachusetts; *A General Catalogue of Trustees, Teachers and Students of Westford Academy, Westford, Massachusetts, 1792–1895. With an Account of the Celebration of the One Hundredth Anniversary* (Westford, Mass., 1912).

12. Edwin R. Hodgman, *History of the Town of Westford in the County of Middlesex, Massachusetts, 1659–1883* (Lowell, Mass.: Morning Mail Co., Printers, Westford Town History Association, 1883), p. 315.

13. Ibid., pp. 325–326; *Annual Report, School Committee, Westford; Catalogue of the Officers and Students of Westford Academy, 1870, 1872, 1874.*

14. Ogilvie and Choquette, "Nettie Maria Stevens," p. 295.

15. Ibid., p. 295.

16. Ibid., pp. 297–298.

17. Rossiter, *Women Scientists in America,* pp. 47–48.

18. Theodor Boveri, "Ueber mehrpolige Mitosen als Mittel zur Analyse des Zellkerns," *Verhandlungen der Physikalischen-medizinischen Gesellschaft zu Würzburg,* 1902, *35:* 67–90, trans. Salome Gluecksohn-Waelsch in B. H. Willier and Jane M. Oppenheimer, eds., *Foundations of Experimental Embryology* (Englewood Cliffs, N.J.: Prentice-Hall, 1964), pp. 76–97.

19. Nettie Maria Stevens, "Further Studies on the Ciliate Infusoria, Licnophora and Boveria" (Ph.D. dissertation, Bryn Mawr College, 1903).

20. Ogilvie and Choquette, "Nettie Maria Stevens," p. 304; see, for example, the following articles by Stevens: "Notes on Regeneration in *Planaria lugubris,*" *Wilhelm Roux' Archiv für Entwicklungsmechanik der Organismen,* 1901, *13:* 396–407; "Regeneration in *Tubularia mesembryantheum,*" *Wilhelm Roux' Archive für Entwick-*

lungsmechanik der Organismen, 1902, *13:* 410–415; "Studies on Ciliate Infusoria," *Proceedings of the California Academy of Sciences,* 1901, *3:* 1–42; "Regeneration in *Antennularia ramosa,*" *Wilhelm Roux' Archive für Entwicklungsmechanik der Organismen,* 1902, *15:* 429–447; "A Histological Study of Regeneration in *Planaria simplissima, Planaria maculata* and *Planaria morgani,*" *Wilhelm Roux' Archiv für Entwicklungsmechanik der Organismen,* 1907, *24:* 350–373; coauthored with Alice Middleton Boring, "Regeneration in *Polychoerus caudatus,*" *The Journal of Experimental Zoology,* 1905, *2:* 335–346.

21. Stevens, "Studies on Ciliate Infusoria"; idem, "Further Studies on the Ciliate Infusoria, *Licnophora* and *Boveria,*" *Archiv für Protistenkunde* 1904, *3:* 1–43.

22. See Garland E. Allen, "Thomas Hunt Morgan and the Problem of Sex Determination, 1903–1910," *Proceedings of the American Philosophical Society,* 1966, *110:* 48–57; Stephen G. Brush, "Nettie M. Stevens and the Discovery of Sex Determination by Chromosomes," *Isis,* 1978, *69:* 163–172.

23. Ogilvie and Choquette, "Nettie Maria Stevens," p. 307; Stevens, "Studies in Spermatogenesis with Especial Reference to the 'Accessory Chromosomes,' " Publication No. 36 (Washington, D.C.: Carnegie Institution 1905).

24. Brush, "Nettie M. Stevens," p. 169; Edmund B. Wilson, "The Chromosomes in Relation to the Determination of Sex in Insects," *Science,* 1905, *22:* 500–502.

25. Thomas Hunt Morgan, "The Scientific Work of Miss N. M. Stevens," *Science,* 1912, *36:* 468–470.

26. Stevens to the secretary of the Carnegie Institution of Washington, 19 July 1903, Archives, Carnegie Institution of Washington.

27. Thomas Hunt Morgan to the Carnegie Institution of Washington, 21 November 1903, Archives, Carnegie Institution of Washington.

28. Martha Carey Thomas to Carnegie Institution of Washington, 30 November 1903, Archives, Carnegie Institution of Washington.

29. Morgan to Carnegie Institution of Washington, 24 March 1904, Archives, Carnegie Institution of Washington; secretary of Carnegie Institution of Washington to Morgan, 26 March 1904, Archives, Carnegie Institution of Washington; secretary of Carnegie Institution of Washington to Stevens, 28 March 1904, Archives, Carnegie Institution of Washington.

30. Ogilvie and Choquette, "Nettie Maria Stevens," p. 301.

31. Stevens to C. B. Davenport, 5 April 1905, Archives, American Philosophical Society.

32. Stevens to Davenport, 7 May 1906; Davenport to Stevens, 9 May 1906; Stevens to Davenport, 13 March 1907; Davenport to Stevens, 14 March 1907; Stevens to Davenport, 17 March 1907, Archives, American Philosophical Society.

33. Mary Brown Bullock, *An American Transplant. The Rockefeller Foundation and Peking Union Medical College* (Berkeley: University of California Press, 1980), p. 2.

34. Alice M. Boring was the third of the four children of Edwin McCurdy Boring and Elizabeth Garrigues Boring. Lydia Truman, the eldest, graduated from Bryn Mawr College and became a teacher of Latin and classics; the second oldest, Katharine, did not graduate from college but became the wife of an educator and

minister, and Edwin, the youngest, obtained a doctorate from Cornell and became a professor of psychology at Harvard.

35. A. M. Boring, "Thomas Hunt Morgan," *Bryn Mawr Alumnae Bulletin* (February 1946).

36. In 1951, when Conklin was president of the American Philosophical Society, he invited Boring to attend the meeting as his personal guest. Since members were allowed only one guest, "usually their wives, but Conklin was then a widower," it was an honor. Edwin Guerigues Boring, "Comment on Clara Woodruff Hull, 'Alice Middleton Boring—1904'."

37. A. M. Boring, application for Graduate School, Bryn Mawr College; idem, "A Study of the Spermatogenesis of the Membracidae, Jassidae, Cercopidae and Fulgoridae," (Ph.D. dissertation, Bryn Mawr College, 1907). Reprinted from *The Journal of Experimental Zoology,* 1907, *4,* 470–533.

38. A. M. Boring, "On the Effect of Different Temperatures on the Size of the Nuclei in the Embryo of Ascaris megalocephala, with Remarks on the Size Relations of the Nuclei of Univalens and Bivalens," *Archiv für Entwickelungsmechanik der Organismen,* 1909 *28:* 118–124; idem, "A Small Chromosome in Ascaris megalocephala," *Archiv zur Zellforschung,* 1909, *4,* 120–131.

39. Boring collaborated with Pearl on several papers. Raymond Pearl and A. M. Boring, "Fat Deposition in the Testis of the Domestic Fowl," *Science,* 1912, *36:* 833–835; idem, "Some Physiological Observations Regarding Plumage Patterns," *Science,* 1914, *39:* 143–144. The interrelationship between Boring, Stevens, and Pearl is made clear in notes published by Boring in which she noted problems in determining whether male or female chickens were homozygous for sex. Pearl began the work, Boring added to it, and then they sent materials to Stevens at Bryn Mawr.

40. See *Bryn Mawr Alumnae Quarterly* (July 1918) for Boring's views of her job at Maine.

41. This commission consisted of Harry Pratt Judson, president of the University of Chicago, Francis W. Peabody of the Harvard Medical School, and Roger S. Greene, then of the U.S. Consular Service. Three commissions in all were involved: the Burton Commission (1909), the First China Medical Commission (1914), and the Second China Medical Commission (1915).

42. Roger S. Greene, "The Rockefeller Foundation in China," *Asia. Journal of the American Asiatic Association,* 1919, *19:* 1117–1124; CMB, Inc., Box 62, Folder 435, Rockefeller Archive Center (hereafter RAC).

43. John Z. Bowers, *Western Medicine in a Chinese Palace. Peking Union Medical College, 1917, 1951* (New York: The Josiah Macy, Jr. Foundation, n.d.), p. 35.

44. *Dedication Ceremonies and Medical Conference, Peking Union Medical College, September 15–22, 1921* (Peking: PUMC, 1922), pp. 63–64.

45. Charles W. Young, "Peking Union Medical College, Requirements," CMB, Inc., Box 126, folder 912, RAC.

46. Marjory Eggleston, secretary to Wallace Buttrick, CMB secretary, 25 July 1918, RAC; Mary E. Ferguson, *China Medical Board and Peking Union Medical College. A Chronicle of Fruitful Collaboration, 1914–1951* (New York: China Medical Board of New York, 1970), p. 232.

47. William R. Hutchison, *Errand to the World. American Protestant Thought and Foreign Missions* (Chicago: University of Chicago Press, 1987).

48. Faculty members, RG 1, Series 601, Box 25, Folder 235, RAC.

49. McLean to A. M. Dunlop, 9 April 1919, CMB, Inc., Box 73, Folder 517, RAC.

50. Report, CMB, Inc., Box 73, Folder 511, RAC.

51. Minutes, Administrative Board, 25 May 1920; 28 May 1920, CMB, Inc., Box 511, RAC.

52. Roger Greene diary, 22 January 1919, RAC; A. M. Boring to E. G. Boring, 18 August 1926, Boring family correspondence, Pusey Archive Collection, Harvard University.

53. Greene, diary, CMB, Inc. Box 62, Folder 437, RAC.

54. John Leighton Stuart, *Fifty Years in China* (New York: Random House, n.d.), p. 50.

55. Boring to Roger S. Greene, 18 October 1922, CMB, Inc., Box 15, Folder 103, RAC.

56. The previous head of the PUMC Biology Department, Charles W. Packard, planned to return to the United States after his tour was completed and, consequently, was not a part of the considerations. Houghton to Greene, 21 November 1922, CMB, Inc., Box 15, Folder 103, RAC.

57. Greene to Henry Houghton, 21 October 1922, CMB, Inc., Box 15, Folder 103, RAC.

58. Boring to J. Leighton Stuart, 11 January 1923, Special Collection Archives, Yale Divinity School.

59. Stuart to Boring, 15 December 1922, Special Collection Archives, Yale Divinity School.

60. Greene to Boring, 24 February 1923, CMB, Inc., Box 15, Folder 103, RAC; Boring to Greene, 4 March 1923, CMB, Inc., Box 15, Folder 103, RAC.

61. Laurence Allen Schneider, "Genetics in Republican China" (unpublished manuscript), pp. 2–3.

62. A. M. Boring to E. G. Boring, 10 July 1925, Boring family correspondence, Pusey Archive Collection, Harvard University. Although Boring was hired on a temporary basis, Stuart wanted her to remain permanently. However, Yenching was committed to hiring qualified Chinese faculty members whenever possible; therefore, the university was hesitant about making a definite commitment to her.

63. Ibid.

64. A. M. Boring to E. G. Boring, 18 August 1925, Boring family correspondence, Pusey Archive Collection, Harvard University.

65. A. M. Boring to Raymond Pearl, 19 August 1929, Library, American Philosophical Society.

66. Pearl to A. M. Boring, 27 September 1929, Library, American Philosophical Society.

67. A. M. Boring to G. K. Noble, 26 February 1929, Dept. of Herpetology, Archives, American Museum of Natural History.

68. A. M. Boring to E. G. Boring, 1 June 1929, Boring family collection, Pusey Archive Collection, Harvard University.

69. A. M. Boring to Noble, 8 October 1929, Dept. of Herpetology, Archives, American Museum of Natural History.

70. Noble to A. M. Boring, 12 November 1929, Dept. of Herpetology, Archives, American Museum of Natural History.

71. William J. Haas, *Botany in Republican China: The Leading Role of Taxonomy* (unpublished manuscript), pp. 22–25.

72. Sven Hedin, Roy Chapman Andrews (Three Asiatic Expeditions), and numerous expeditions sponsored by the Smithsonian Institution added to the taxonomic knowledge of China.

73. Donna Haraway, "Teddy Bear Patriarch: Taxidermy in the Garden of Eden, New York City, 1908–1936," *Social Text,* 1984, *5:* 21–64, on p. 57.

74. Roy Chapman Andrews, *The New Conquest of Central Asia; a Narrative of the Explorations of the Central Asiatic Expeditions in Mongolia and China, 1921–1930,* vol. 1 (New York: American Museum of Natural History, 1932); Clifford Hillhouse Pope, *The Reptiles of China: Turtles, Crocodilians, Snakes, Lizards,* vol. 10 (New York: American Museum of Natural History, 1935).

75. A. M. Boring, "Early Days of the Peking Natural History Society," *Peking Natural History Society Bulletin,* 1950, *18:* 77–79.

76. A. M. Boring and C. C. Liu, "A New Species of Kaloula with a Discussion of the Genus in China," *Peking Natural History Bulletin,* 1931–1932, *6,* 19–24, on pp. 21–22.

77. A. M. Boring to "Friends," 26 March 1939, Special Collection Archives, Yale Divinity School; Boring to "Friends," 7 May 1939, Special Collection Archives, Yale Divinity School.

78. Pope and A. M. Boring, "A Survey of Chinese Amphibia," *Peking Natural History Society Bulletin,* 1940, *15:* 13.

79. Ogilvie, *Women in Science,* p. 45.

80. Ibid.

81. Ibid.

Léo F. Laporte

4

George G. Simpson, Paleontology, and the Expansion of Biology

The geneticists tended to consider that paleontology was incapable of rising above pure description and they did not even take the trouble to study descriptive paleontology for its bearing on genetics. It was easier to conclude that it had no such bearing. The paleontologists were, as a rule, quite willing to accept this stultifying conclusion, which also spared them the trouble of learning genetics.[1]

Paleontology at the turn of the twentieth century was something of an orphan within the biological sciences. Because of its emphasis on the older biological tradition of comparative morphology and the use of fossils by geologists in determining relative ages of rocks, paleontology had no home in the new institutional centers established for biology, nor was it considered a formal discipline within biology. Worse yet, it did not demonstrate "much likelihood of becoming a foundation for serious programs of research in biology."[2] The Department of Vertebrate Paleontology at the American Museum of Natural History provided the single exception to this otherwise bleak situation. Henry Fairfield Osborn, William King Gregory, and William Diller Matthew of the Department of Paleontology were among the few paleontologists of this era who consistently addressed biological questions in their study of fossils.

During the subsequent twentieth-century expansion of biology, paleontology continued to exclude itself from biology's expanding theoretical and observational base, especially the genetics and ecology of living populations of organisms, until the development of the "evolutionary synthesis" in the late 1930s and early 1940s. Given the American Museum's earlier history as an outpost of biologically oriented paleontology, it is not surprising that another, younger American Museum scientist, George Gaylord Simpson, catalyzed the reconciliation of paleontology with contemporary biology. As the above quotation suggests, until this reconciliation came about, paleontology and biology were at cross-purposes in their search for evolutionary mechanisms. Paleontologists had argued for internally directed evolutionary forces inferred from their fossils that biologists could not corroborate in the laboratory or the field, whereas biologists had focused on problems that seemed irrelevant to the paleontologist. Bones-in-stones were a world apart from flies-in-bottles.

This essay will examine Simpson's role as a prime mover in bringing paleontology and biology closer together during the evolutionary synthesis, thereby resulting in a mutual expansion of the two disciplines: genetics and ecology providing a theoretical context for the actual historical patterns documented by fossils. George Gaylord Simpson (1902–1984) dominated American paleontology for some five decades spanning the middle of the twentieth century. This dominance was both quantitative and qualitative, for not only did Simpson publish hundreds of articles, monographs, and books (his bibliography includes more than 750 entries), but his work had a major impact on contemporary views of the origin, evolution, and classification of mammals; the concepts of historical biogeography; the principles of taxonomy and systematics; biostatistical methods; as well as the formulation of the modern evolutionary synthesis.

Because his research data ultimately derived from fossils in layered sequences of sedimentary rocks, Simpson necessarily straddled the two disciplines of geology and biology. Consequently, Simpson had to be competent in both. His mastery of geology was accomplished by undergraduate and graduate matriculation in geology at Yale University. His education in biology was somewhat adventitious, yet it turned out to be crucial in his intellectual development, especially with respect to his subsequent contributions to evolutionary theory. Both by formal graduate study at Yale and professional association at the American Museum, Simpson realized early in his career the special value of the biological sciences in studying and interpreting vertebrate fossil remains.

This deep appreciation and thorough understanding of the subject enabled Simpson to bring paleontology more fully into the mainstream of biology during the decade of the development of the modern evolutionary synthesis. However, for all Simpson's realization of the importance of biology for paleontology, he was nevertheless equally convinced of the uniqueness of paleontology, particularly in terms of its raw data of fossils-in-rocks and its much longer temporal context, thereby justifying paleontology's own intellectual autonomy.

Biology and Fossils at Yale

> *Biology did interest me, and I began making up [at Yale] the serious gap on this side of my knowledge requisite for paleontological work.*[3]

In 1922 Simpson transferred from the University of Colorado to Yale University for his senior year of college work. Reasons for his transfer included getting what he perceived as the best education in paleontology then available.[4] At Colorado, Simpson had a year-long course in paleontology, but no biology.[5] Yale's liberal arts curriculum required a year of biology, so Simpson enrolled in the two-semester class taught by Lorande L. Woodruff, who founded and taught the basic biology course and whose textbook, *Foundations of Biology,* had just been published. "Woodruff gave a good introduction to Mendelian genetics as of the 1920s. As to evolution, he adhered to what he called 'clarified Darwinism,' accepting natural selection as the only natural explanation of adaptation, but believing that many 'variations are neutral to selection . . . and hinting . . . at . . . some vitalistic factor.[6] Further insight about Woodruff's views on evolution at this time are reflected in an essay on biology that Woodruff wrote for his edited volume, *The Development of the Sciences:* "Today no representative biologist questions the fact of evolution . . . though in regard to the factors there is much difference of opinion. It may well be that we shall have reason to depart widely from Darwin's interpretation of the effective principles at work in the origin of species."[7]

Simpson continued at Yale as a graduate student in the Department of Geology studying vertebrate paleontology. Because biological form and function were considered important by the paleontology faculty, as important as chemistry and physics was for other geologists, the department required its paleontologists to take courses in zoology.[8] Thus Simpson took classes in vertebrate morphology with Ross Harrison, embryology with G. A. Baitsell, and comparative anatomy of vertebrates with W. W. Swingle.[9] Simpson has said about Harrison that he "taught morphology in a rather nineteenth century Teutonic way, very thorough and wonderful basic knowledge, with no emphasis on evolutionary theory."[10]

Simpson's dissertation adviser at Yale, Richard Swan Lull (1867–1957), was himself trained as a biologist. Lull had bachelor and master's degrees in zoology from Rutgers, before taking his Ph.D. in vertebrate paleontology at Columbia under Henry Fairfield Osborn, who was also trained as a biologist. Lull's biological background was clearly evident in his textbook, *Organic Evolution,* which Simpson read over the summer of 1923 before entering graduate school.[11]

The 1929 revision was issued after Simpson left Yale, but it indicates the approach Lull took to the subject when Simpson was at Yale just a few years before. Throughout his book, Lull integrated contemporary biology with the vertebrate fossil record. In Part II, entitled "The Mechanism of Evolution," Lull had seven separate chapters on genetic variation and mutation, heredity, and artificial, sexual, and natural selection as well as inheritance of acquired characters, and orthogenesis and kinetogenesis. In the rest of the book, dealing specifically with particular fossil groups, Lull included as much biology of living organisms as was appropriate to understand fossils. For each fossil group he discussed its place in nature, form and function, habitat and habit, particular specializations, and ontogenetic as well as phylogenetic history.

Although the book had obvious Lamarckian and vitalistic overtones, Simpson was certainly taken with Lull's paleobiological approach. According to Simpson, Lull infused "life into the bones . . . the whole fauna, flora, and landscape of the distant past."[12] Simpson later remarked that "Lull was considered a great authority on evolutionary theory—and he was, in the sense of being well informed on the ideas, old and new, in this field. He never had an original idea of his own about theories of evolution, but just expounded everyone's views as if all were equal. That was very useful to me only in telling me what the established alternatives were and what to read."[13] So both by formal education and by the example of his dissertation adviser, Simpson learned early in his career the importance of biology in thinking about fossils.

These particular biological influences with respect to the biology of living organisms or evolutionary theory were not especially apparent in Simpson's dissertation and postdoctoral work on Mesozoic mammals.[14] Like most paleontological research of the era, these latter studies emphasized descriptive morphology, taxonomy and systematics, and phylogenetic interpretation. However, before he reached age twenty-four, Simpson wrote two predissertation articles on Mesozoic mammals that did explicitly consider fossils as once-living. In the first, which he called "a study in paleobiology, an attempt to consider a very ancient and long extinct group of animals not as bits of broken bone but as flesh and blood beings," Simpson interpreted the multituberculates—an extinct, yet longest-lived mammalian order that had worldwide distribution and considerable diversity—as "living animals," using the skull and dentition of the marsupial rat-kangaroo as a living analogue to infer a herbivorous diet of "cycads, cones and nuts, and angiosperm fruits and seeds." Morphological analysis of the limbs of one multituberculate genus in light of the locomotor function of extant animals led Simpson to the further conclusion that the extinct form was "probably a swift moving and agile quadruped."[15]

In the second article, Simpson reconstructed the "schema of the paleoëcological relationships of the terrestrial and aquatic cenobiota" of the Late Jurassic Morrison Formation of Wyoming, showing the inferred ecosystem of the fossils—including invertebrates, fishes, and reptiles as well as mammals.[16] As Simpson himself remarked years later, this was among the first—if not *the* first—such application of the ecological concepts of trophic levels and food webs to a fossil association, which was not to become commonplace for several more decades.[17] While the specific content of both these studies was original for the time, their spirit certainly is anticipated in Lull's approach to fossils as developed in his text *Organic Evolution*.

Biology and Fossils at the American Museum

> *I have a debt, a loyalty to the museum; the best place for me to do what I wanted to do.*[18]

After finishing his Yale doctoral dissertation on North American Mesozoic mammals, Simpson took a year of postdoctoral work at the British Museum to expand his study to include British and Continental mammalian fossils. He returned from England in the fall of 1927 to take up a position as assistant curator in the Department of Vertebrate Paleontology at the American Museum of Natural History in New York. This move was to have enormous implications for the rest of Simpson's career. The museum's collections were outstanding, providing Simpson with new fossil mammals to work on and extensive specimens of living species for reference, as well as a superb research library close at hand. His prescribed duties, too—at the beginning at least—were sufficiently minimal that he could spend most of his time on his own research. The museum was even more important intellectually for Simpson in that his senior colleagues included William King Gregory and Henry Fairfield Osborn, both of whom always approached fossils as once-living, complex biological systems. William Diller Matthew, another distinguished paleontologist, whose position at the American Museum Simpson filled in 1927, played an even earlier role in Simpson's life, when Simpson was Matthew's young field assistant on a museum expedition to Texas in 1924. In his first year in graduate school at Yale, Simpson had applied to the American Museum for summer work, either as a preparator of fossils in the laboratory or as a field assistant.[19]

In his autobiography, Simpson said of William Diller Matthew (1871–1930) that he was "trained by Matthew at least as much as by my formal professor, Lull."[20] In a posthumously published, personal memoir, first written at the time of Matthew's death in 1930, Simpson gives some details about Matthew's influence on him that hot Texas summer.[21] Simpson recalled that Matthew revealed "his life's dream (one never gratified) to build up an exhibition of vertebrate paleontology . . . to show not so much the zoologic classification as the grand history of life, the succession of fauna & the associations of animals at one time & in one place."

One might suppose that Matthew was confiding to Simpson in the field what he would subsequently tell his Berkeley students: "You cannot understand or appreciate the past history of life without knowing something about animals and plants, about their structure and mechanism and habits and how they are classified and arranged—what they are and where they live and how they live."[22] Matthew also had strong thoughts about what paleontologists could contribute to evolutionary biology. In his 1922 presidential address to the Paleontological Society he remarked, "I do not agree with a distinguished Columbia professor [Thomas Hunt Morgan] who declared not long ago [in 1916] that paleontologists had no business to reason or draw conclusions from their specimens, but should content themselves with describing and illustrating them. . . . Paleontologists, with the facts before them as to what actually did take place in the evolution of a race of animals, may claim the right to reason and draw conclusions from these data as to the methods and causes of the transmutation of the species."[23] Likewise, Matthew's interests in historical biogeography, as reflected in his classic *Climate and Evolution,* also influenced Simpson's own extensive subsequent research in the past geographic distribution of terrestrial mammals.[24]

William King Gregory (1876–1970) had become a formal member of the scientific staff at the American Museum in 1911, after taking his Ph.D. at Columbia under Henry Fairfield Osborn, although he had been Osborn's assistant for a dozen years before that. Gregory's influence on Simpson is evident from Simpson's memorial to Gregory.[25] Gregory's scientific interests were very broad, and several overlapped with those of Simpson, including mammalian classification, Mesozoic mammals, evolution of mammalian dentition, and functional anatomy. Some flavor of Gregory's scientific philosophy that particularly impressed Simpson is captured in the latter's words: "Gregory was a pioneer in the study of the anatomy of parts of animals, both recent and fossil, not merely in a descriptive way but with primary consideration for their functions in the lives of whole, living organisms." Simpson did not formally acknowledge this debt to Gregory in any of his publications, but he did point out in his memorial to Gregory the significance of Gregory's "palimpsest theory" of differing rates of evolution of individual characters and character complexes—Gavin de Beer's "mosaic evolution" in present-day terms.[26] This theory may have contributed to Simpson's own views about differential evolutionary rates within a given lineage that he developed so elaborately in *Tempo and Mode in Evolution.*[27]

Perhaps the single, most specific, catalytic effect Gregory had on Simpson's career (that we can document) was Simpson's preparation of a paper "following a suggestion by Professor Gregory" for a symposium that he was organizing for the American Society of Naturalists in 1936 on "supra-specific variation," that is, differentiation and evolution above the species level.[28] Much later in life Simpson called this one of his two "door-opening papers" that led to his more explicit consideration of the principles of classification and evolutionary theory.[29] At the time, however, when he sent the manuscript to Gregory, Simpson noted that "it has never been entirely clear to me exactly what sort of paper you wanted, and I may have missed the point altogether. This is in any case an unusually labored effort and for some reason I have had great difficulty in selecting and organizing, from the embarrassingly vast amount of data at hand."[30] Whatever ambivalence he may have had at the time, the ideas introduced in the paper about "tempo"—or rates—and "mode"—or styles—of evolution were to assume central importance in his 1944 classic.[31]

Not the least important influence on the young Simpson was Henry Fairfield Osborn (1857–1935), who was virtually synonymous with the American Museum in the early decades of this century. Osborn's training in various biological disciplines, including histology under William H. Welch in New York, embryology under Francis M. Balfour at Cambridge, comparative anatomy under Thomas H. Huxley in London, and neuroanatomy at Princeton, as well as having collected Eocene mammals in Wyoming and studied Mesozoic mammals in England resulted in his communicating "a soundness of training and a passion for research that have had a profound effect on vertebrate paleontology."[32] Owing to his intellectual and material resources as well as his energetic and dynamic leadership, Osborn created an institution of singular distinction. Within vertebrate paleontology, Osborn developed a department in which "problems of biogeography, the process and pattern of evolution, or the relationship between inheritance and development, more than the traditional issues in systematics and stratigraphy, were the heart and soul."[33]

Osborn thus gave warrant to paleontologists to pursue theoretical issues, which Simpson did not hesitate to acknowledge. Shortly after publication of *Tempo and Mode,* Simpson reviewed "some of the historical background of modern evolutionary theories," including especially his own. Because "none of the mechanistic schools . . . seemed fully adequate to interpret paleontological observations . . . [and] paleontologists themselves failed to find a satisfying alternative . . . [,] some paleontologists began to support a non-mechanistic, that is, in a broad sense, a vitalistic explanation." Simpson noted that an outstanding example of such a nonmechanistic explanation was Osborn's theory of aristogenesis, and that "it was almost inevitable that such an escape from the dilemma would be sought by a deeply philosophical paleontologist of his period."[34] Osborn provided a forceful example to Simpson of what the important questions were in paleontology. In his biographical sketch of Osborn, Simpson repeated Osborn's own words to the effect that he "always found the mere assemblage of facts an extremely painful and self-denying process . . . [whereas] the discovery of new principles is the chief end of research."[35]

It was into this professional ambience that Simpson stepped, at age twenty-five, when he became an assistant curator in the Department of Vertebrate Paleontology at the American Museum.

Simpson's Biological Insights of the 1930s

> *Now age thirty-five and more than ten years out of graduate school I dared to take modest personal steps in the direction of principles and theory.*[36]

In the mid-1930s, Simpson worked on several projects that were seminal for virtually all his subsequent work. In spring of 1936 Simpson completed a monograph describing Paleocene mammals of Montana, which included brief discussions of the paleoecology of the faunas, their postmortem preservation (today's "taphonomy"), and the value of univariate statistics in aiding taxonomic decisions.[37] Later that year Simpson delivered two papers, one that addressed general problems of the evolutionary interpretation of fossils, and a second that specifically illustrated how such interpretations can be made using the extinct mammalian order Notoungulata.[38] During the same time Simpson was also collaborating with the psychologist Anne Roe (whom he married in 1938) on a book for zoologists and paleontologists explaining how statistics were the "best means of describing and interpreting what animals are and do."[39] More generally, throughout the 1930s, Simpson was working on a new classification of mammals that took a unified approach to the analysis and interpretation of living and extinct animals, as well as informing himself of the new developments in population genetics.[40] It was his reading in 1937 of Theodosius Dobzhansky's *Genetics and the Origin of Species* that crystallized all his previous training and thinking about biological paleontology and encouraged him to take modest steps in the direction of principles and theory. "When I read [Dobzhansky] it opened a whole new vista to me of really explaining the things that one could see going on in the fossil record and also by study of recent animals. I began pulling together into this framework but also with a good many points that were not involved in the work of the geneticist, [I] began thinking of what my own by this time rather long studies in the history of life might mean within this context."[41]

The bulk of the monograph on Paleocene mammals contained standard morphological description and taxonomic discussion of fossils that Simpson and others had collected over the years in the Fort Union Formation of Montana, However, the point of view taken by Simpson was rather original for such a paleontological monograph. Simpson used elementary statistics for describing mammalian teeth: their individual lengths and widths, means and ranges of observations, standard deviations, and coefficients of variability as well as the *t*-test for comparing samples. The purpose of these calculations was to assist him in making objective and reproducible inferences about the probable limits of variation acceptable in defining species. As Simpson noted, "most other writers on the question of species making in paleontology have insisted on making due allowance for variation, but they have adduced no real, objective criteria as to what 'due allowance' should be." Rather than just paying lip service to the new concept of species as populations of breeding individuals rather than as idealized types, Simpson was advocating that "the methods of statistics provide the desired means of measuring variation accurately and the necessary criterion as to whether this variation is . . . of the sort normal in a species."[42]

Simpson used these statistical results, for example, to justify his inference that there were, indeed, two similar but statistically distinct species of carnivore that lived in geographically different areas of the region, thereby confirming the principle of "ecological incompatibility." Simpson observed that the principle had been qualitatively applied previously by W. D. Matthew.[43] But here, Simpson was invoking the principle using a quantitative, statistical test to support what biologists would later refer to as Gause's hypothesis of "competitive exclusion."[44]

In the paper "Patterns of Phyletic Evolution," Simpson recommended the use of the population concept of species, noting that the effect of "paleontological psychology has been to direct attention, to a disabling extent, to the individual specimen, and the number of publications on fossil mammals in which a suite of specimens has been correctly studied as representative of a group can be almost counted on the fingers of one hand."[45] Using a small hypothetical set of data, comprising a range of values for a given morphologic character that increase in dimension through a sequence of rock strata, Simpson indicated the three sorts of qualitative interpretation that might result. The data might record the decline over time of one smaller species and the rise of another larger one, or possibly just one species in which the character was increasing through time, or perhaps even the gradual divergence of one larger species from the smaller ancestral one. Simpson then explained the basis for discriminating among the three alternative interpretations by simple statistical analysis of the data, stratum-by-stratum, through the rock sequence. Simpson declared years later that this paper "marked his abandonment of the typological thinking of my college teachers and started aiming me toward statistical biometry and the deeper investigation of evolutionary theory and taxonomic stance."[46]

A third seminal paper of this period by Simpson was the one on supraspecific variation that leads to evolutionary diversification above the species level; as noted above, this was a topic suggested to him by W. K. Gregory. In the first half of the paper, Simpson reviewed the history of an extinct group of hoofed herbivores, the Notoungulates, almost entirely confined to South America during the Tertiary Period. Simpson demonstrated that rates of evolution can vary markedly within different characters of an otherwise homogeneous taxon, leading to distinct higher categories (e.g., different suborders within the order Notoungulata). He showed that close parallel evolution is often manifest between otherwise distinctly different taxa (e.g., similar limb elongation, toe reduction, and increase in molar height both in Notoungulates and in true horses). Finally, Simpson inferred that there is inherent variability within interbreeding populations whose subsequent segregation can eventually produce supraspecific taxa (e.g., evolution of distinct lineages having characteristic types of molars from a highly variable ancestral species whose variability spans that of the separate, descendant lineages). The second half of the paper critiqued an earlier paper by Alfred C. Kinsey by "adding what seems to be the best or most generally held paleontological opinion on each concept" regarding various principles for the recognition of higher taxa that Kinsey had himself debated.[47] What was significant about this discussion was that a paleontologist, using fossils as evidence, was prepared to argue about biological principles and concepts on equal terms with a biologist.

Simpson made the same point in *Quantitative Zoology* by recommending that paleontological samples be given the same statistical analysis one should use for biological samples. As implied by the title and illustrated by examples throughout the text, Simpson intentionally blurred the distinction between biological data and paleontological data, and the conclusions one could draw from them. For him there was no important qualitative difference between living and fossil: "Paleontological mensuration differs little from that of the hard parts of recent animals."[48] Therefore, theoretical interpretations based on fossils were, in principle at least, as sound and cogent as those relying on living materials. Whatever else one might be concerned with about the size and quality of a given sample—fossil or alive—both kinds of samples had similar validity, all other things being equal.

Simpson further conjoined biology and paleontology in his work as curator at the American Museum. There he had to identify, label, sort, and catalog fossil mammalian specimens. Simpson soon developed a set of concepts and principles to relate them to living taxa so he could physically organize the museum's collections. Like any workable filing system, Simpson had to set up a catalog that had some rhyme and reason to it.[49] Progressively elaborated over the next two decades, Simpson formulated a comprehensive approach to taxonomy and systematics. This resulted in an outline classification of mammals (1931), followed by a monograph that expanded the previous outline and included a discussion of the principles of classification (1945). Still later he produced a full-blown exposition of the principles of animal taxonomy (1961). In all three works, Simpson observed no formal distinctions between living and fossil.[50] On the contrary, every piece of information, whatever its source and disciplinary jurisdiction, was useful: "The data of neozoology are highly pertinent to the problems of phylogeny and major classification, but this work has fallen more and more into the field of the palaeozoologist who should, for this purpose, be a competent general zoologist as well as a palaeontologist. . . . In so difficult a study [as taxonomic classification] it is inexcusable to reject offhand any line of evidence that might give light."[51]

Simpson thus threw down the gauntlet in 1937 when he asserted "that there is no natural barrier between genetic and paleontological research and that both must eventually unite in any final synthesis of modes of evolution."[52] Now fully prepared to undertake a more comprehensive treatment of the subjects he had been exploring these several years, Simpson picked up the gauntlet himself in the spring of 1938 when he began work on *Tempo and Mode in Evolution*.

Tempo and Mode in Evolution

> *The present purpose is to discuss the "how" and . . . not the "what" . . . [the how] is more immediately interesting to the nonpaleontological evolutionist.*[53]

Simpson's *Tempo and Mode in Evolution* is one of the half-dozen or so books that are the pillars of the modern evolutionary synthesis; it also helped bring paleontology back into the mainstream of biological science.[54] Hence it is tempting for paleontologists today to see this book and its author sui generis, like Venus rising from the sea, without historical roots or intellectual genealogy. But as Ronald Rainger cogently argues and the previous sections of this essay suggest, that view is historically inaccurate.[55] The genesis of *Tempo and Mode* was itself evolutionary, not revolutionary. By his previous training and the examples of his mentors, Simpson recognized the importance of the larger biological issues raised by his fossil materials, and he had learned the skills to address those issues. He was no mere describer and namer of fossil bones and teeth. In fact, as a paleontological treatise, what was left out of *Tempo and Mode* was as unexpected as what was included. There was very little discussion of specific fossil taxa per se: whether their detailed morphological description; their temporal progression, geologic period by geologic period; or their phylogenetic sequence, from ancestor to descendant.[56]

Simpson's task in *Tempo and Mode in Evolution* was to demonstrate that what happens on the individual level of populations of living organisms—genetic and phenotypic variation, natural selection, differential survival and reproduction, acclimatization and adaption, migration, and so on—is necessary and sufficient to explain the much longer-term, morphological transformations of skeletal hard-parts seen in the fossil record. Thus Simpson proposed to show how the microevolution of population genetics of organisms, viewed within an ecological context of those organisms interacting with their diversified environments, could explain the macroevolution documented by paleontology. That is, he was projecting the three-dimensional concepts of biology into the fourth, temporal, dimension of paleontology. As he phrased it a decade later, "I am trying to pursue a science that . . . has no name: the science of four-dimensional biology or of time and life. Fossils are pertinent . . . but *Drosophila* is equally pertinent."[57]

As a way of encapsulating his argument, Simpson developed the concept of the "adaptive grid," a field of discrete, noncontinuous ecological zones of increasingly finer grain. When considered on a time-scale ranging from ecological (microevolutionary) to geological (macroevolutionary) time, populations might become adapted, early on, within a narrow zone with little subsequent change (very slow evolution, or Simpson's "bradytely"); if the prospective adaptive zone is wide enough, populations might diversify and become increasingly specialized within various subzones (slow to fast evolution, or "horotely"); or rarely, populations might jump the gap between major adaptive zones and radiate rapidly into a new adaptive zone altogether (very fast evolution, or "tachytely"). What was so original about *Tempo and Mode* was that it expressly addressed the "how" of the evolutionary transformations familiar to the paleontologist in terms that fit well with experimental genetics and field biology of living organisms.

Of course to get to the point where he could discuss the "how," he had first to demonstrate that fossils do record highly varying rates of evolution (Chapter I); that population genetics explained the mechanisms of evolution on the ecological time-scale (II); that large, systematic gaps in the fossil record were due to particularly high rates of evolution (III) whereas very low rates of evolution of so-called living fossils could be attributed to "survival of the unspecialized" in unchanging, long-lived habitats (IV); and that, therefore, all intrinsic evolutionary mechanisms, such as orthogenesis, inertia and momentum, and racial senescence, hypothesized by earlier paleontologists, were incorrect (V).

These chapters served dual purposes by addressing two different audiences. On the one hand, they made accessible to paleontologists the recent discoveries of population genetics and introduced important ecological concepts. For example, how the rapid shift might have occurred in the Oligocene epoch when some browsing horses "jumped the gap" to a grazing way of life as certain favorable genetic variants underwent strong selection. On the other hand, these chapters were meant to convince the biologists that the fossil record, indeed, had useful content that could make genuine contributions to evolutionary theory. For instance, Simpson demonstrated how one could measure relative and absolute rates of evolution, either in terms of changes in morphology within lineages (like fossil horse molars) or in terms of taxonomic categories between fossil lineages (like horses and marine cephalopods).[58] Simpson thus began the process of winning over the two factions, who had had a history of talking at cross-purposes, when they talked to each other at all.

Having brought the two camps together, Simpson could next develop how interactions between "organism and environment" (Chapter VI), expressed generally but not always through natural selection, led to adaptation. Depending on the temporal and ecologic scales used, various outcomes ensued, whether new geographic races, or subspecies ("microevolution"), new species and genera ("macroevolution"), or new families, orders, or higher taxa ("megaevolution").[59] Thus one easily moved along a continuum from the realm of the population geneticist and field biologist to that of the paleontologist. The final chapter of *Tempo and Mode* (VII) summarized the full argument of the book by relating specific patterns of evolution ("mode") to their complementary rates of evolution ("tempo") against the backdrop of environmental context.

Response to *Tempo and Mode*

> *I am consoled by the conviction that* [Tempo and Mode] *had some historic value . . . and that its thesis has stood up well.*[60]

Simpson devoted three chapters of *Tempo and Mode* to explicitly biological subjects: the "determinants of evolution," "organism and environment," and how they are integrated in varying "modes of evolution." Thus half of the book was "evolutionary biology," which illuminated what Simpson called the "how" of evolution, and in the other half "phylogenetic examples [from fossils] are introduced as evidence and to give reality to the theoretical discussion [based on genetics]." Intentionally or not, Simpson thus gave equal time to the two disciplines he was trying to integrate, biology and paleontology. But the initial response to the book was by no means equally enthusiastic in both camps.

Reviews by biologists were quite lavish in their praise: "from now on no competent discussion of the mechanisms of macroevolution can be made outside the ambit of Simpson's analysis"; "hereafter no treatise on the causes and modes of evolution can ignore" *Tempo and Mode;* the "most important contribution yet to come from paleontology on the methods by which evolution takes place"; Simpson "has performed the double task of reminding neozoologists of many facts of paleontology which they have tended to overlook as unfamiliar or inconvenient, and at the same time showing the possibility of accounting for them in genetic terms"; and "the reconciliation [of genetics] with paleontology."[61] Of course, almost every reviewer had the usual, minor reservations about one thing or another, but the overall impact was definitely one of unanimous approval of Simpson's book by the biological community.[62]

No doubt part of this approval from biologists was self-congratulatory, because Simpson fully accepted and integrated data and interpretations of population geneticists into his reading of the fossil record. But surely it was more than self-interest, because by extending the explanatory power of the new genetics not only to a different set of data but to data viewed at completely different temporal and ecological scales of reference, *Tempo and Mode* made the genetic argument more consilient. Simpson by this time had been elected to the American Philosophical Society and the National Academy of Sciences, so biologists could only be pleased that a leading student of paleontology validated their results and expanded their realm of explanation.

The paleontological community, however, paid little attention at first to *Tempo and Mode*. For example, it was not reviewed by the leading North American paleontological journal. The only paleontologist to take note of it was Glenn Jepsen, ironically, for the *American Midland Naturalist*. Jepsen praised "its services as a debunking device for some traditional and stereotyped opinions" of paleontologists, and noted the enthusiasm of biologists, but he pointed out aspects of it that they ignored: namely "the most original contributions to interpretation of fact and epistemology . . . [that is,] the expansion and correlation of scientific reasoning in paleontology."[63]

Why the lack of immediate response by the paleontologists? It wasn't because they were too obtuse to understand the argument, but rather that the argument did not directly involve their own day-to-day research. As Rainger has noted, there was a strong tradition in paleontology in the late nineteenth and early twentieth centuries that concerned itself chiefly with empirical morphological description, and its change and sequence.[64] Although by the 1930s there was some theoretical work by people like Gregory, Matthew, and Osborn (as noted above), the great majority of practitioners were still not so theoretically inclined. So while paleontologists accepted the overall thesis of *Tempo and Mode,* there was not an obvious display of it in their published writings.

Even Simpson remarked upon this lack of response several years after the book appeared. He tabulated the subject matter of "principal papers" in the *Journal of Paleontology* for the years 1939 and 1949, that is, before and after *Tempo and Mode*. On the basis of fifty-eight papers published in 1939 and sixty papers in 1949, he noted virtually no difference in either the absolute numbers or the relative proportions of articles devoted to "descriptive morphology and systematics": 71 percent for both years. "Mainly geological" papers accounted for 17 percent versus 11 percent, whereas those "mainly biological" in content, 8 percent versus 7 percent. Most surprising of all, within the "mainly biological" category, there was just a single paper on "evolutionary theory" published in 1939 and *none* in 1949, and just a handful in either year concerned with "principles of systematics, phylogeny, and functional morphology."[65]

Simpson introduced this analysis by noting that it is "the opinion of some paleontologists that there has been some change in the scope and emphasis of paleontological research"—presumably owing to the formulation of the evolutionary synthesis—such that "descriptive work would tend to broaden, to emphasize increasingly the bearing of the fossils described on their geological and biological settings . . . and perhaps give more attention to such subjects as ecology, functional and broadly comparative morphology, [and] evolutionary processes." After tabulating his statistics, Simpson concluded that the "figures certainly reveal no trend away from straight descriptive studies," and he asked rhetorically, "does this [lack of change] betoken desirable stability and maturity of research programs or does it indicate lack of progress and undesirably narrow, routine, and unimaginative approaches to research?"[66]

Because scientific journals obviously are supported by people who have a vested interest in what the journal publishes, they tend to resist change in editorial policy. In fact, Simpson already had sensed this resistance and therefore was instrumental in founding the journal *Evolution,* first published in 1947 by the fledgling Society for the Study of Evolution, whose members spanned the diverse spectrum of specialties unified by the evolutionary synthesis. As the first president of that society in 1946, Simpson raised the money to get the journal started.[67] In the early decades of *Evolution*'s existence, a number of paleontologists—including Simpson—contributed articles dealing with "evolutionary theory." Their contribution proportionally waned somewhat in later years, if only because the number of such journals interested in "theory" had since increased.

Only a few contemporaries of Simpson paid more than lip service to the synthesis, but full-scale incorporation of the results of the evolutionary synthesis into a "paleontological tradition" required the subsequent education of a new generation of paleontologists in microevolutionary theory—especially genetics and ecology—as well as the new macroevolutionary theory, rather than a making-over of those already established in the discipline.[68] Everett C. Olson has recalled that as editor of *Evolution* in the 1950s, he invited Alfred S. Romer (1894–1973), one of America's leading vertebrate paleontologists, to "please write us some articles," but Romer demurred, saying, "But I don't write that kind of evolution."[69] Nevertheless, by the mid-1960s, two decades after the consolidation of the evolutionary synthesis (and the publication of *Tempo and Mode*), paleontology had indeed broadened its scope and emphasis, symbolized by common usage of the word "paleobiology," to distinguish the new practitioners from the old.

Adoption of the Population Concept of Species

> *A change was then [1939] in the air, especially as regards systematics which among all the ramifications of zoology necessarily remains its basic discipline. . . . The population approach has now [1960] become usual in systematics and has spread into all branches of zoology.*[70]

Although *Tempo and Mode,* and more generally the evolutionary synthesis, was slow in having an impact on the ongoing, day-to-day practice of paleontology, there was at least the passive effect among paleontologists of abandoning the older, now outmoded concepts. As Ernst Mayr has noted, the evolutionary synthesis refuted a "number of misconceptions that had the greatest impact on evolutionary biology. This includes soft inheritance, saltationism, evolutionary essentialism, and autogenetic theories."[71] Simpson's work endorsed hard inheritance; claimed that evolutionary transformation was gradual (in terms of the degree of genetic changes from one generation to the next, although the rate of change could be quite variable, from very slow to very fast), and thus that any apparent gaps among fossils were artifacts of the record; debunked inherent factors in driving evolution (like racial senescence, momentum, inertia, and orthogenesis); and accepted an antiessentialist, populational view of the species. *Tempo and Mode* thus was very important in catalyzing the end of one conceptual era in paleontology and in ushering in another, even if the transition did not occur overnight.

However, there was one specific area where there was fairly immediate, active response, and that was the shift from the typological to the population concept of species. As quoted above, Simpson had remarked about the disabling effects of the attention paid to the individual fossil specimen by paleontologists. Prior to the evolutionary synthesis, species were commonly defined on the basis of a single tooth or bone or shell that provided the "type," exemplifying the sort of Platonic "essentialism" noted by Mayr. During the writing of *Tempo and Mode,* Simpson addressed this particular issue in a separate article, by proposing the concept of the "hypodigm" to replace the type: "All the specimens used by the author of a [new] species as his basis for inference, and this should mean all the specimens that he referred to the species, constitute his hypodigm of that species. . . . The hypodigm is a sample from which the characters of a population are to be inferred."[72]

As two of his distinguished paleontological contemporaries observed some time later: "In earlier years the all important specimen of the species was the type: a sort of enthroned little god, in the image of which all other individuals of the species were supposed to have been made. We are coming to quite a different position now . . . what we must consider is the population as a whole—the hypodigm to use Simpson's term."[73] "This whole scheme [of taxonomic nomenclature] collapsed like a house of cards when George Simpson published his short but epoch-making paper on *Types in Modern Taxonomy*. Simpson emphasized that in an interbreeding population each individual is as much as any other a part of the species. Therefore a true conceptional image of the species—which he called the hypodigm—must encompass the extremes as well as the means of variation within such a population."[74]

This conversion to populational thinking, of course, was more than cosmetic, because any serious thinking about the natural populations of fossil organisms led to broader considerations of genetic variation, selection, interactions of organism and environment—in short, many of those elements included within the evolutionary synthesis. So even if the practical significance of *Tempo and Mode* might initially have seemed somewhat remote to paleontologists, in fact this new perspective on more ordinary descriptive morphology and taxonomic work continuously reinforced *Tempo and Mode*'s biological concepts. It emphasized for paleontologists that the theoretical contributions biology was making to paleontology could be reciprocated because fossils had to validate biological theory. Simpson observed that whereas "early paleontologists regularly cited the fossil record as evidence *against* the reality of evolution . . . [now] no theory of evolution can long be satisfactory, even to the geneticists and systematists, unless it is explicitly shown to be harmonious with the factual record of evolution as revealed by paleontology. Moreover, there are very essential parts of a general theory of evolution that cannot be based on the study of recent animals and plants, alone."[75]

Ernst Mayr, too, acknowledged this role for paleontology: "The study of long-term evolutionary phenomena is the domain of the paleontologist. He investigates rates and trends of evolution in time and is interested in the origin of new classes, phyla, and other higher categories. Evolution means change and yet it is only the paleontologist among all biologists who can properly study the time dimension. If the fossil record were not available, many evolutionary problems could not be solved; indeed, many of them would not even be apparent."[76]

In one important aspect of the populational species concept, Simpson did indeed distance himself from the biologists. As an evolutionist interested in Darwinian "descent with modification," yet constrained as a paleontologist in observing that modification always in terms of morphological transformation of hard parts (bones, teeth, and shells) through a sequence of sedimentary rocks, Simpson recognized a distinction between evolutionary species and biological species. "An evolutionary species is a lineage (an ancestral-descendant sequence of populations) evolving separately from others and with its own unitary evolutionary role and tendencies." The biological concept of species as given by Simpson followed Mayr's definition: "groups of actually or potentially interbreeding natural populations, which are reproductively isolated from other such groups."[77] Simpson, the paleontologist, was preoccupied with which species gave rise (over time) to the next species, whereas Mayr, the biologist, would be preoccupied with physical isolation (across geographic area) giving rise to new species. Thus Simpson's species concept emphasized the vertical or temporal aspect of the species whereas Mayr's definition emphasized the horizontal or geographic aspect. Although both concepts of species are consistent with each other, the differences in emphasis reflected the different sorts of data available to the field biologist and the paleontologist—the former working within a virtually zero-time dimension of the present across fine-grained geographic space, the latter working within a greatly expanded temporal dimension across much coarser-grained geography. Given the nature of the fossil record, it is therefore extremely difficult for paleontologists to resolve time and space with sufficient precision to document Mayrian speciation the way biologists can infer it. Clearly, then, it is the longer-term, larger-scale changes and trends in hard-part morphology that will more often occupy paleontologists. However much they might be interested in observing "fossilized speciation," the inherent nature of the fossil record usually makes it inaccessible or impracticable.

Disciplinary Independence for Paleontology

> *Paleontology is characterized, but not fully defined, by having its own objective subject matter: fossils. Fossils occur in rocks, and they are organisms. Their extended study necessarily overlaps widely into both of the broader (or more miscellaneous) sciences of geology and biology.*[78]

For all of Simpson's awareness and acknowledgment of the importance that biology had for paleontology, he was at the same time equally cognizant that the study of fossils was a unique and separate discipline. The source of this disciplinary identity was the nature of the raw, basic data: fossils in rocks. It is "the flow of new discoveries and data from the field, laboratory preparation of specimens, and study of their morphology and taxonomy" that provide the "most basic essentials for continued progress in vertebrate paleontology." But however important such data and study were—what Simpson was fond of calling the "what" of paleontology—he also believed it important to "consider problems of broader and more theoretical biological interpretation that arise after the basic data, taxonomic and geologic, are in hand"—what he called the "how" of the science.[79]

Simpson was so adamant about maintaining disciplinary identity for paleontology that he considered resigning from the American Museum when the new director, Albert Parr, disbanded the Department of Vertebrate Paleontology in June 1942 and transferred Simpson and his two colleagues to other biological departments.[80] Parr thought evolution was "a finished issue" and "would contribute little further knowledge of any importance."[81] Obviously, such an opinion was anathema to Simpson: "A whole new chapter in the history of evolution theory is just beginning. It is almost incredible that the Museum, with its great tradition of interest in evolutionary studies, is not taking a more leading part in this work." But before Simpson could decide whether to resign or not, his enlistment into the Army was accepted. By the time he returned from military service two years later, Parr had reversed his plan. Simpson then became chairman of the new Department of Geology and Paleontology, where he reigned until 1958 when he had another disagreement with director Parr.[82]

A more positive example of Simpson's belief in paleontology's mission was his role in the founding of the Society of Vertebrate Paleontology (SVP) in the early 1940s. Early in this century vertebrate paleontologists had their own professional society, but in 1907 decided to join with the somewhat more numerous invertebrate paleontologists to create the Paleontological Society.[83] Almost immediately, however, the Paleontological Society became a section of the still larger Geological Society of America. Soon vertebrate paleontologists, always a small coterie of scientists, were greatly outnumbered by increasing numbers of invertebrate paleontologists and geologists. Gradually, vertebrate paleontologists found that their interests at national meetings were given low priority. Initially, they tried to remedy their loss of identity by becoming a formal section of the Paleontological Society, but eventually this too was unsatisfactory. After several years of discussion, the SVP became formally organized in 1941, with a membership of some 150 vertebrate paleontologists.

Simpson's role in the founding of the SVP was central. He was secretary-treasurer during its provisional first year in 1940 and its elected president in its first year of formal existence. An archival photograph, taken of the organizational meeting of the SVP at Harvard, shows Simpson (secretary-treasurer) sharing the dais with Alfred Sherwood Romer (president), another doyen of twentieth-century American vertebrate paleontology, looking out over two dozen vertebrate paleontologists gathered below. This photograph is a synecdoche for the two dominant strains within paleontology: the "how" of the theoretician (Simpson) juxtaposed with the "what" of the comparative anatomist (Romer).

During 1941, the organizational year of the SVP, Simpson published several small issues of a mimeographed "news bulletin," which contained not only the usual "who, what, when, and where," but also an essay by him entitled "Some Recent Trends and Problems of Vertebrate Paleontological

Research." In 1941 Simpson was also in the middle of writing *Tempo and Mode* and his monograph on the classification of mammals. So we might expect a plea for broader, more theoretical consideration of the meaning of the basic raw data of vertebrate paleontology in such an essay. On the contrary, the essay is entirely concerned with the increasing decentralization of vertebrate fossil collections, such that it was more and more difficult for researchers to have easy access to them. Moreover, lack of adequate preparation, curation, and library resources aggravated the problems. "This society is a proper forum for their discussion and could perhaps be a means toward their solution. . . . It is certainly practical to improve the conditions for research without retarding, indeed while facilitating, the broader distribution of collecting, exhibition, and public education."[84]

Given Simpson's fundamental contributions to the evolutionary synthesis and to other theoretical subjects, one might easily overlook his extensive writings on descriptive morphology, classical taxonomy, and systematics. A recent tabulation of Simpson's oeuvre counts 109 titles and 6,675 pages of his work as theoretical and synthetic, and 224 titles and 5,785 pages as empirical collection-oriented.[85] Obviously, in word and deed, Simpson was as committed to the "what" of vertebrate paleontology as to the "how." Accordingly, while he saw the importance of biology for illuminating the "how," he also was firmly convinced that the "what" deserved full recognition on its own terms as a separate discipline of paleontology. It therefore could not be subsumed under biology—or geology for that matter. Consequently, however strongly he believed that paleontology should draw upon observations, principles, and concepts within the biological sciences, he was equally convinced that all such theorizing must be firmly grounded in the fossils themselves and their geologic context.

Conclusion

> *I have collected a great many fossils, described even more, and named a good number of them. . . . Beyond that, I have taken a broader stance and a more theoretical and subjective one always in part in geology but increasingly also in organismal and evolutionary biology.*[86]

When Simpson entered paleontology as a neophyte in the 1920s, the discipline was estranged from the rest of biology because of the apparent contradiction between putative mechanisms of evolution endorsed by biologists and the evolutionary history of organisms as interpreted by paleontologists. This estrangement, in turn, was a continuation of the more general, ongoing dichotomy between the older morphological tradition of the nineteenth century represented by paleontology and the newer experimental fields, like genetics and ecology, that were expanding within biology in the twentieth century.

During the 1930s and 1940s, Simpson demonstrated that the contradiction between paleontology and biology was only apparent and that in fact the two disciplines were mutually supportive of each other's conclusions. Simpson was thus instrumental in expanding contemporary biology's theory into the deep time represented by fossil history. This accomplishment resulted from his training in biology by distinguished teachers as well as from day-to-day institutional contact with a few paleontologists already sympathetic to what biology had to offer their discipline. The mutual relationship of the two disciplines, each expanded by the other, in the latter half of the twentieth century thus reflects in no small measure the contributions of George Gaylord Simpson. His election to the presidencies of the American Society of Mammalogists (1962), the Society for Systematic Zoology (1962), and the American Society of Zoologists (1964) was obvious acknowledgment of his contributions to biology, as was the award of more than a dozen international medals and prizes by scientific societies and organizations.[87]

Acknowledgments

I thank Ronald Rainger for his instruction in the history of modern paleontology as well as Philip Gingerich, David Webb, and the Friday Harbor conferees for their helpful critical readings of an early draft of this essay.

Notes

1. George G. Simpson, "Tempo and Mode in Evolution," *New York Academy of Sciences Transactions,* 1946 *2nd Ser. 8:* 45–60, on p. 53. First presented as a lecture in November 1945 in the Section of Biology, and its title was the same as the book Simpson published the previous year, which is considered one of the half-dozen founding works of the evolutionary synthesis. Simpson noted in his introduction to this lecture that the book attempted "to take some of the data available to paleontologists and not to geneticists, and to relate these to recent developments in evolutionary theory," whereas the lecture focused on the "historical background of modern evolutionary theories" and the divergence and subsequent convergence of paleontology and genetics on this subject. (I refer to this article in subsequent citations as "N. Y. Academy of Science article" to distinguish it from Simpson's *Tempo and Mode* book.)

2. Ronald Rainger, "Vertebrate Paleontology as Biology: Henry Fairfield Osborn and the American Museum of Natural History," in Ronald Rainger, Keith Benson, and Jane Maienschein, eds., *The American Development of Biology* (Philadelphia: University of Pennsylvania Press, 1988), pp. 219–256, on p. 220. For development of the position of paleontology vis-à-vis biology around the turn of the twentieth century, see also Ronald Rainger, "The Continuation of the Morphological Tradition: American Paleontology, 1880–1910," *J. Hist. Biol.,* 1981, *14:* 129–158;

idem, "What's the Use: William King Gregory and the Functional Morphology of Fossil Vertebrates," *J. Hist. Biol.,* 1989, *22:* 103–139. For a review of the state of paleontology just before the evolutionary synthesis, see Stephen J. Gould, "G. G. Simpson, Paleontology, and the Modern Synthesis," in Ernst Mayr and William B. Provine, eds., *The Evolutionary Synthesis* (Cambridge, Mass.: Harvard University Press, 1980), pp. 153–172; esp. pp. 153–157; Léo F. Laporte, "Simpson's *Tempo and Mode in Evolution* Revisited," *American Philosophical Society Proceedings,* 1983, *127:* 365–417; esp. pp. 373–377; and Simpson, N. Y. Academy of Sciences article, pp. 45–54.

3. George G. Simpson, handwritten autobiographical notes, 1933, unpaginated; typed and revised in 1954; handwritten additions to typescript in 1970, Archives, American Philosophical Society.

4. George G. Simpson, *Concession to the Improbable* (New Haven: Yale University Press, 1978), p. 16. For more biographical background, especially letters from his early adult life, see G. G. Simpson, *Simple Curiosity: Letters from George Gaylord Simpson to His Family, 1921–1970,* Léo F. Laporte, ed. (Berkeley and Los Angeles: University of California Press, 1987).

5. University of Colorado undergraduate transcript.

6. See Ernst Mayr's abridged and edited answers that Simpson gave to Mayr's questionnaire dealing with his role in the evolutionary synthesis, in Mayr and Provine, eds., *The Evolutionary Synthesis,* p. 453; see also Lorande L. Woodruff, *Foundations of Biology* (New York: Macmillan, 1922), p. 378.

7. Lorande L. Woodruff, "Biology," in L. L. Woodruff, ed., *The Development of the Sciences* (New Haven: Yale University Press, 1923), p. 258.

8. Interview with George G. Simpson by Léo F. Laporte, 3 February 1979, Tucson, Arizona.

9. Yale graduate transcript.

10. Mayr and Provine, *Evolutionary Synthesis,* p. 453.

11. Yale graduate transcript; Richard S. Lull, *Organic Evolution,* revised ed. (New York: Macmillan, 1929).

12. George G. Simpson, "Memorial to Richard Swan Lull," *Geological Society of America Proceedings,* 1958, *Annual Report for 1957,* pp. 127–134, on p. 130.

13. Mayr and Provine, *Evolutionary Synthesis,* p. 453.

14. George G. Simpson, *A Catalogue of the Mesozoic Mammalia in the Geological Department of the British Museum* (London: British Museum [Natural History], 1928); and idem, "American Mesozoic Mammalia," *Memoirs of the Peabody Museum of Natural History of Yale University,* 1929, *3:* pt. 1, pp. 1–235.

15. George G. Simpson, "Mesozoic Mammalia. IV. The Multituberculates as Living Animals," *American Journal of Science,* 1926, *5th ser. 11:* 228–250, on p. 228, 246, and 247.

16. George G. Simpson, "The Fauna of Quarry Nine," *American Journal of Science,* 1926, *5th ser. 12:* 1–16.

17. George G. Simpson, *Why and How: Some Problems and Methods in Historical Biology* (Oxford and New York: Pergamon Press, 1980), p. 86.

18. George G. Simpson interview with Léo F. Laporte, 18 August 1981, Tucson.

19. Léo F. Laporte, "George G. Simpson (1902–1984): Getting Started in the Summer of 1924," *Earth Sciences History,* 1990, *9:* 62–73.

20. Simpson, *Concession,* p. 34.

21. George G. Simpson, "G. G. Simpson's Reflections on W. D. Matthew," *Palaios,* 1986, *1:* 200– 204.

22. William D. Matthew, *Outline and General Principles of the History of Life,* University of California Syllabus Series, No. 213 (Berkeley: University of California Press, 1928), on p. 6.

23. William D. Matthew, "Recent Progress and Trends in Vertebrate Paleontology," *Annual Report of the Smithsonian Institution for 1923* (Washington, D.C.: U.S. Government Printing Office, 1925), pp. 273–289.

24. William D. Matthew, "Climate and Evolution," *Annals of the New York Academy of Sciences,* 1915, *24:* 171–318; see Léo F. Laporte, "Wrong for the Right Reasons: George G. Simpson and Continental Drift," *Geological Society of America, Centennial Publication 1,* 1985, pp. 273–285 for a discussion of why Simpson's biogeographic theory seemingly led him astray with respect to Alfred Wegener's hypothesis for continental drift.

25. George G. Simpson, "William King Gregory, 1876–1970," *American Journal of Physical Anthropology,* 1971, *35:* 155–173.

26. Ibid., pp. 158–160. The notion is that a given kind of organism "is a mosaic of primitive and advanced characters, of generalized and special features," owing to differing evolution of its various components; Ernst Mayr, *Animal Species and Evolution* (Cambridge, Mass.: Harvard University Press, 1963), p. 598.

27. George G. Simpson, *Tempo and Mode in Evolution* (New York: Columbia University Press, 1944).

28. George G. Simpson, "Supra-specific Variation in Nature and in Classification," *American Naturalist,* 1937, *71:* 236–267, on p. 236. The venue for Simpson's "door-opener" was most appropriate. As noted on the title page, the paper was read 30 December 1936 in Atlantic City, N.J. at a symposium of the American Society of Naturalists in joint session with the American Society of Zoologists, the Botanical Society of America, the Genetics Society of America, the American Phytopathological Society, and the Ecological Society of America. This was precisely the audience for "presenting the paleontological view-point on the zoological problem of higher [taxonomic] categories" (ibid., on p. 236).

29. Simpson, *Concession,* pp. 80–82.

30. George G. Simpson to William K. Gregory, 16 November 1936; Folder G, Archives, American Philosophical Society.

31. Laporte, "Simpson's *Tempo and Mode* Revisited," pp. 367–369.

32. George G. Simpson, "Henry Fairfield Osborn," *Dictionary of American Biography,* 1944, *11* (suppl. 1): 584–587, on p. 585.

33. Rainger, "Vertebrate paleontology as Biology," p. 221.

34. Simpson, N. Y. Academy of Sciences article, p. 52. "Aristogenesis" was the name Osborn gave to the inherent drive within organisms toward future perfection. Simpson's use of "philosophical" here was to indicate that Osborn was often seeking to address in his scientific writings the larger issues of "how and why" with respect to organic evolution, not just merely the "what." In this colloquial sense Simpson, too, was "philosophical." See Léo F. Laporte, "The World into Which Darwin Led Simpson," *J. Hist. Biol.,* 1990, *23:* 499–516.

35. Simpson, *Dictionary of American Biography,* p. 586.

36. Simpson, *Concession,* p. 81. Why the ten-year gap (1926–1936) between graduate school and Simpson's first "theoretical" papers? Recall that is wasn't until the early 1930s that the work of the population geneticists, like Fisher, Haldane, and Wright, was beginning to have an impact, and not until 1937 did Simpson discover them through Dobzhansky's classic. See n. 40.

37. George G. Simpson, "The Fort Union of the Crazy Mountain Field, Montana and Its Mammalian Faunas," *United States National Museum Bulletin,* 1937, *169:* 1–287.

38. George G. Simpson, "Patterns of Phyletic Evolution," *Geological Society of America Bulletin,* 1937, *48:* 303–314; idem, "Supra-Specific Variation.

39. George G. Simpson and Anne Roe, *Quantitative Zoology* (New York: McGraw-Hill, 1939), p. vii.

40. Key writings of this time that Simpson cited in his 1944 book, *Tempo and Mode in Evolution,* were R. A. Fisher, *The Genetical Theory of Natural Selection* (Oxford: Clarendon Press, 1930); J. B. S. Haldane, *Causes of Evolution* (New York and London: Harper, 1932), and various articles by Sewell Wright, especially "Evolution in Mendelian Populations," *Genetics,* 1931, 16: 97–159; and Theodosius Dobzhansky's *Genetics and the Origin of Species* (New York: Columbia University Press, 1937). In tape-recorded comments he made in 1975, providing background on his various published works, Simpson acknowledged the important influence of these writings on his own research as well as Dobzhansky's *Genetics and the Origin of Species.* See George G. Simpson, "Transcription of Comments on His Bibliography," 1975, Archives, American Philosophical Society. There is no evidence that Dobzhansky personally influenced Simpson in one-on-one conversations during this period. Both did later overlap in time in New York City during 1940–1942; Dobzhansky came to Columbia from the California Institute of Technology in 1940, and then Simpson went to war in 1942. However, Simpson did not yet have an adjunct appointment at Columbia—that wasn't until 1945—and he had a reputation for *not* engaging in conversation on these issues, even at the museum. As Mayr has remarked, "we [Simpson and Mayr] never talked about these things. We sat at the same lunch table at the museum and never once discussed evolutionary theory." (Ernst Mayr interview with Léo F. Laporte, 22 October 1980, Cambridge, Mass.)

41. Simpson, tape-recorded comments, p. 45. See also Simpson's comments in Mayr and Provine, *The Evolution Synthesis,* p. 456. How accurate are such post-hoc memories, some forty years later? It does seem that Dobzhansky's book was the key turning point in Simpson's appreciation of genetics for his paleontological research, but he probably was not as explicitly aware of the work of the geneticists until then, because he does not refer to any of that research in these three 1937 articles (two delivered orally in 1936). Thereafter in many of his publications, besides *Tempo and Mode,* he does cite not only Dobzhansky's 1937 book but also the earlier genetic work on which it was based and to which it referred.

42. Simpson, "Fort Union of the Crazy Mountain Field," p. 63.

43. Ibid., p. 64; William D. Matthew, "Range and Limitations of Species as Seen in Fossil Mammal Faunas," *Geological Society of America Bulletin,* 1930, *41:* 271–274.

44. The hypothesis is that organisms closely similar in morphology are for that reason inferred to be closely similar in ways of life as well, so that interspecific

competition would eventually drive one species out; by living in separate geographic areas they would no longer compete with one another.

45. Simpson, "Patterns of Phyletic Evolution," pp. 308–309.

46. Simpson, *Why and How,* p. 112. "Typological" because the species is defined on the basis of single, type specimen to which all subsequent specimens are referred for inclusion or exclusion with respect to the named species. For species named according to a "population" concept the reference point is a sample of several specimens, usually not identical, and therefore the species represents some combination of all the specimens. However, by the rules of zoological nomenclature, one specimen must still be designated the "type specimen," which becomes the physical identity bearing the new species name. The important difference is that "typological thinking" views the type specimen as a manifestation of some idealized, unvarying species; whereas "population thinking" allows for, even expects, variation among individuals drawn from an interbreeding population, which is, of course, the basic unit of evolution.

47. Simpson, "Supra-Specific Variation," p. 250. Alfred C. Kinsey (1894–1956) did extensive research on the life history and evolution of gall wasps in Mexico and Central America; later in his career he became well known to the general public for his studies on human sexual behavior.

48. Simpson and Roe, *Quantitative Zoology,* p. 23.

49. George G. Simpson interview with Léo F. Laporte, 2 February 1979, Tucson.

50. George G. Simpson, "A New Classification of Mammals," *American Museum of Natural History Bulletin,* 1931, *59:* 259–293; idem, "The Principles of Classification and a Classification of Mammals," *American Museum of Natural History Bulletin,* 1945, *85:* 1–350; idem, *Principles of Animal Taxonomy* (New York: Columbia University Press, 1961).

51. Simpson, "Principles of Classification," pp. 2, 7.

52. Simpson, "Supra-Specific Variation," p. 250.

53. Simpson, *Tempo and Mode,* p. xviii.

54. Gould, "Simpson, Paleontology, and the Modern Synthesis," pp. 36 ff.; Ernst Mayr, *The Growth of Biological Thought* (Cambridge, Mass.: Harvard University Press, 1982), pp. 568, 607 ff.; Laporte, "*Tempo and Mode* Revisited," pp. 402 ff. *Tempo and Mode* was "written at intervals between the spring of 1938 and the summer of 1942"; George G. Simpson, *Major Features of Evolution* (New York: Columbia University Press, 1953), p. ix.

55. Ronald Rainger, "Just Before Simpson: William Diller Matthew's Understanding of Evolution," *American Philosophical Society Proceedings,* 1986, *130:* 453–474; and Rainger, "Vertebrate Paleontology as Biology."

56. See Gould, "Simpson, Paleontology, and the Modern Synthesis," pp. 153–172, for a discussion of what was unique about Simpson's treatise and the role it played in the evolutionary synthesis, and Laporte, "*Tempo and Mode* Revisited" for a more detailed discussion of the specific content of *Tempo and Mode* as well as the state of paleontology just before its publication, how the book was received, and its current status.

57. Simpson, *Major Features of Evolution,* p. xii.

58. Simpson, *Tempo and Mode,* pp. 92–93 and pp. 8 ff.

59. Although Simpson made these three distinctions, he abandoned them in his later discussions, preferring to follow geneticist Richard Goldschmidt's usage "that historical changes within species be called microevolution while those from species upward are to be called macroevolution." Simpson made this distinction for convenience only, and noted that it coincided with a difference in the domain of materials available for direct study to the experimental biologist and paleontologist, respectively, and not to qualitative differences between the two kinds of evolution. See Simpson, *Major Features of Evolution,* pp. 338– 340.

60. George G. Simpson, "The Compleat Paleontologist?" *Annual Reviews of Earth and Planetary Science,* 1976, *4:* 1–13, on p. 5.

61. Theodosius Dobzhansky, "Genetics of Macro-Evolution; a Review of Tempo and Mode in Evolution," *Journal of Heredity,* 1945, 36: 113–115, on p. 114; Bentley Glass, "Review of *Tempo and Mode in Evolution, "Quarterly Review of Biology,* 1945, *20:* 261–263, on p. 261; G. Evelyn Hutchinson, "Review of *Tempo and Mode in Evolution," American Journal of Science,* 1944, *243:* 356–358, on p. 356; Julian Huxley, "Genetics and Major Evolutionary Change: Review of *Tempo and Mode in Evolution," Nature,* 1945, *156:* 3–4, on p. 3; Sewell Wright, "A Critical Review," *Ecology,* 1945, *26:* 415–419, on p. 415.

62. See Laporte, "*Tempo and Mode* Revisited," pp. 404–405, for fuller discussion of the contemporary reviews of *Tempo and Mode.*

63. Glenn L. Jepsen, "Review of *Tempo and Mode in Evolution," American Midland Naturalist,* 1946, *35:* 538–541 on p. 538. Glenn L. Jepsen (1903–1974) was a contemporary of Simpson's, a Princeton professor of vertebrate paleontology, and, like him, a student of early Cenozoic mammals. Because of their closeness in age and field of research, Jepsen was impressed by, yet somewhat envious of, Simpson's greater professional renown.

64. Rainger, "The Continuation of the Morphological Tradition."

65. George G. Simpson, "Trends in Research and the Journal of Paleontology," *Journal of Paleontology,* 1950, *24:* 498–499.

66. Ibid., p. 498 and p. 499.

67. Simpson, *Concession,* p. 129.

68. Several such American contemporaries of Simpson's who were consistently theoretically inclined were Preston E. Cloud (b. 1912) and Norman D. Newell (b. 1909) among the invertebrate paleontologists and Everett C. Olson (b. 1910) and Glenn Jepsen among vertebrate paleontologists; all were sufficiently prominent to have been elected to the National Academy of Sciences.

69. Everett C. Olson, transcription of comments made at the second workshop on the founding of the evolutionary synthesis, 11–12 October 1974, Cambridge, Mass., p. 27, Archives, American Philosophical Society.

70. George G. Simpson, Anne Roe, and Richard Lewontin, *Quantitative Zoology,* 2nd ed. (New York: McGraw-Hill, 1960), p. v. Simpson defined systematics as "the scientific study of the kinds and diversity of organisms and of any and all relationships among them," whereas "taxonomy is the theoretical study of classification, including its bases, principles, procedures, and rules." Simpson, *Principles of Animal Taxonomy,* pp. 7, 11.

71. Mayr, *Growth of Biological Thought,* p. 570.

72. George G. Simpson, "Types in Modern Taxonomy," *American Journal of Science,* 1940, *238:* 413–431, on p. 418.

73. Alfred S. Romer, "Vertebrate Paleontology, 1908–1958," *Journal of Paleontology,* 1959, *33:* 915–925, on p. 919. Romer at that very time was about to offer Simpson the Alexander Agassiz professorship at Harvard's Museum of Comparative Zoology, of which he was the director.

74. Carl O. Dunbar, "A Half Century of Paleontology," *Journal of Paleontology,* 1959, *33:* 909–914, on p. 911. Dunbar (1891–1970) was professor of invertebrate paleontology and stratigraphy at Yale and director of its Peabody Museum of Natural History.

75. Simpson, N. Y. Academy of Sciences article, pp. 49, 57; emphasis in the original.

76. Mayr, *Animal Species and Evolution,* p. 11.

77. Simpson, *Principles of Animal Taxonomy,* p. 153, and on p. 150.

78. George G. Simpson, "Some Problems of Vertebrate Paleontology," *Science,* 1961, *133:* 1679–1689, on p. 1679.

79. Ibid., p. 1679 and p. 1680.

80. John M. Kennedy, "Philanthropy and Science in New York City: The American Museum of Natural History," (Ph.D. dissertation, Yale University, 1969), pp. 242–243; George G. Simpson interview with Léo F. Laporte, 3 February 1979, Tucson.

81. Quoted in Kennedy, "Philanthropy and Science," p. 241, and letter from Simpson to museum trustee, quoted in ibid., p. 246.

82. This latter disagreement was administrative, not disciplinary. Parr pressured Simpson to retire from the chairmanship, because of his slow recovery from a near-fatal accident in Brazil. Simpson not only retired the chairmanship but left the following year when offered an Alexander Agassiz Professorship by A. S. Romer, director of Harvard's Museum of Comparative Zoology.

83. This summary of the history of the SVP follows George G. Simpson, "History of the Society and Its Predecessors," *News Bulletin,* Society of Vertebrate Paleontology, March 1941, 1, pp. 1–3. See also John A. Wilson, "The Society of Vertebrate Paleontology 1940–1990. A Fifty Year Retrospective," *Journal of Vertebrate Paleontology,* 1990, *10:* 1–39.

84. George G. Simpson, "Some Recent Trends and Problems of Vertebrate Paleontological Research," pt. 1, *News Bulletin,* Society of Vertebrate Paleontology, 1941, no. 3, pp. 2–4; pt. 2, 1941, no. 4, pp. 10–11.

85. Philip D. Gingerich, 1986, "George G. Simpson: Empirical Theoretician," *Contributions to Geology,* University of Wyoming Special Paper 3, p. 3–9.

86. Simpson, *Concession,* p. 268.

87. Simpson was mostly ill when he was president of the ASZ; he and his wife, Anne, suffered "his and her heart failures and were hospitalized" over the summer, and they spent most of the rest of the year on a convalescent South Seas cruise (Simpson, *Concession,* pp. 214–215). President-elect Theodore Bullock became de facto ASZ president during 1964, before assuming his own formal presidency in 1965.

Adele E. Clarke

5

Embryology and the Rise of American Reproductive Sciences, circa 1910–1940

Not a single English-language book on reproductive sciences was published until F. H. A. Marshall's *Physiology of Reproduction* appeared in Britain in 1910.[1] Yet by 1940, reproductive sciences had both emerged and coalesced as a line of work in American biology, medicine, and agriculture. Investigators had garnered well over a million dollars in external research funding from major foundations through support from a prestigious National Research Council committee. And preeminence in reproductive sciences had shifted from British and European centers to the United States.[2]

Why, how, and from whence did American reproductive sciences emerge after 1910? One of the striking features of reproductive science is the lateness of its development. The physiologies of each of the other major organ systems—neurological, cardiovascular, digestive, and respiratory—were already the focus of considerable scientific attention by the late nineteenth century.[3] Why were reproductive sciences so "late"?

This essay argues that in the United States until about 1910, biological problems of reproduction were confounded in a nexus of concerns about heredity, development, and evolution. Not until these concerns were clarified into the disciplines of genetics and developmental embryology did boundaries of a third discipline, reproductive science, also begin to emerge. Most histories of turn-of-the-century life sciences attend carefully to the emergence of genetics and reframing of developmental embryology.[4] In contrast, I am asserting that there was a three-way splitting of emergent problem structures and that reproductive science was the third emergent discipline.

Like American genetics, American reproductive sciences emerged largely from embryology. This was not the case in Britain, where medically and agriculturally oriented physiologists initiated study of reproductive phenomena.[5] Nor was it the case in Germany and Austria, where gynecologists pioneered in reproductive investigations, or in France, where clinical and agricultural concerns were primary.[6] Rather, the unique configuration of the limited numbers of American biologists and other life scientists, their institutional situations, and the centrality of embryology to American concerns (perhaps as a viable, shared meeting ground on which to build American biology)[7] all contributed to making embryology the field of origin of both American genetics and reproductive sciences.

It was not, of course, only the three-way split that led to the development of reproductive sciences. Beginning in 1889, new theories of hormonal regulation of bodily function gained acceptance.[8] The concept of blood-borne "chemical messengers" was central to understanding many reproductive phenomena. Yet intellectual forces alone cannot explain why research on reproduction not only got under way in the United States by the 1920s, but by the 1930s had become the centerpiece of a major committee of the National Research Council: the Committee for Research on Problems of Sex.[9] A variety of altered social and economic conditions paved the way for development of reproductive investigations in biology, medicine, and agriculture after the turn of the century, countering the impropriety of such work due to its association with sexuality and reproduction.

This essay first examines evidence for the three-way split and the emergence of reproductive sciences from embryology. Next is a brief review of hormonal research up to the eve of World War I, setting the stage for a detailed examination of two key sets of embryological investigations: one undertaken by Frank Rattray Lillie of the University of Chicago, and a second set by George Papanicolaou, a zoologist then working under the sponsorship of Charles Stockard at Cornell Medical College. Both studies were published in 1917, and both were fundamental to the meteoric rise of American reproductive sciences between the wars. Finally, returning to issues of the three-way split, I discuss why a significant number of outstanding American investigators, almost all embryologists, then became the first generation of American reproductive scientists. The conclusions focus on the importance of the three-way split to understanding the expansion of American biology in the twentieth century.

The Three-Way Split: Genetics, Developmental Embryology, and Reproductive Sciences

During the late nineteenth and early twentieth centuries, there was a complexly interwoven nexus of scientific problems one might inelegantly term "heredity-evolution-reproduction-development." Problem structures as well as both disciplinary and professional boundaries were relatively undifferentiated and porous at this time. "Rediscovery" of Mendel's work in 1900 and clarification of a sex-related chromosome permitted a new distinction to be made between the genetics of sex and the biology of reproduction. This initiated the splitting of the nexus of concerns into differentiated problem structures that then became the bases of emergent disciplines. One of the major literatures in the history of the life sciences focuses on these issues, emphasizing the emergence of genetics.[10] But this ambitious literature has ignored reproductive processes, which were part of the nexus of problems various investigators were dividing into new disciplines early this century.

During the decade after 1900, genetics emerged with a problem structure focused largely on genes and chromosomes and with a definition of heredity largely restricted to the sexual transmission of heritable factors from one generation to the next. As Frederick B. Churchill has noted, "Given this orientation, which stressed sexual events on both the organism and cellular level, it is not surprising that when the geneticist takes stock of his historical past, he stresses those discoveries and advances that elucidated the nature of sexuality."[11] But as cytology joined Mendelism, attention was almost exclusively focused on hybridization experiments and concern over the factorial constitution of the gametes—genes. This conceptualization eliminated the problems of both naturalists and biometricians from the core activity of such genetics[12] and excluded the mechanisms leading to cellular differentiation—the traditional embryological problem of development.[13] Geneticists left problems of growth and generation to embryologists.[14] This web of problems, usually focused through organ systems, remained the jurisdiction of the newly bounded developmental and experimental embryology that now essentially ignored former evolutionary and regenerative concerns.[15]

Thus new problem areas were clarified. Embryologists could delineate epigenetic development from the organized egg. Cytologists understood that chromosomes maintain structural individuality and could explore sub-chromosomal morphological "factors." Geneticists could focus on the nature, function, and effects of inherited factors and demonstrate their location on chromosomes.[16]

My argument is that investigators interested in problems of reproduction also had a much clearer agenda for action and potential new constituencies for reproductive biological research. In 1910, F. H. A. Marshall defined reproduction as a research area with its own problem structure in his classic volume, *The Physiology of Reproduction.* Marshall stated in his introduction: "The all-important questions of heredity and variation, although intimately connected with the study of reproduction, are not here touched upon, excepting for the merest reference, since these subjects have been dealt with in various recent works, and any attempt to include them would have involved the writing of a far larger book. Similarly, the subject matter of cytology, as treated in such works as Professor Wilson's volume on the cell, is also for the most part excluded."[17]

Marshall thus argued that the scope of the physiology of reproduction constituted a distinctive set of scientific problems without heredity, variation (evolution), and cytology. Moreover, these problems deserved a volume devoted exclusively to them. Marshall thus astutely grasped the splitting of emergent disciplines so characteristic of his day, and marked the boundaries of a new discipline from the perspective of a reproductive investigator.[18]

Frank Rattray Lillie, an embryologist then turning to reproductive investigations, offered further argument for attending to these phenomena in his 1919 text, *Problems of Fertilization:* "It is commonly said that there are *two* main problems in the physiology of fertilization, viz.: the initiation of development, or activation, and biparental inheritance. Indeed, so long as we regard fertilization primarily as a problem of *prospective* significance in the life of the organism, *we shall miss the more specific aspects of the process*. Once fertilization is accomplished, development and inheritance may be left to look after themselves."[19]

Lillie's call for a retrospective view of fertilization can be viewed as an assertion of a third set of problems in the physiology of fertilization: sexual reproduction systems. That subject was the basis of the third line of work delineating a three-way split: What must happen before fertilization can itself take place? How are the products joined in fertilization themselves produced? What other processes are retrospectively involved in fertilization?[20]

Three years later, Lillie presented "A Classification of Subjects in the Biology of Sex" to the recently formed National Research Council Committee for Research in Problems of Sex. Lillie's framework was much broader than Marshall's and included: (1) Genetics of Sex; (2) Determination of Sex; (3) Sex Development and Differentiation; (4) Problem of Sex Interrelations (sexuality), (5) Sex Functions (physiology of reproduction); and (6) Systematics of Sex in Animals and Plants.[21]

While the initial boundaries among several of these lines of work were blurred,[22] three distinctive lines of scientific work were gradually delineated and clarified during the first three decades of the twentieth century: genetics, developmental embryology, and reproductive sciences. As Philip J. Pauly has noted, biololgists were moving into new areas as part of broader attempts to expand their constituencies, once they had secured positions in the academy.[23] Reproductive sciences were part of this expansion.

The Emergence of Endocrinology

The emergence of modern endocrinology between 1890 and 1905 was central to the emergence of reproductive sciences, for many reproductive processes are hormonally controlled. The scientific import of discoveries of hormones lay in the shift from neurological theories and explanations as triggers of physiological processes to biochemical explanations. This was, of course, linked to the rise of biochemistry. In 1895 Edward Schafer (later Sir Edward Sharpey-Schafer) became a champion of the theory of internal secretions; for him, endocrinology was the basis of what he termed the "New [Chemical] Physiology."[24] In 1905, the term "hormone" was introduced by Ernest Starling.[25] Gradually, endocrinology became a specialty, splitting off from general physiology in a similar fashion to the splitting of genetics from embryology. While reproductive endocrinology was part of general endocrinology, reproductive sciences (including reproductive endocrinology) also involved an array of nonendocrinological problems that had a somewhat distinctive identity.[26]

Modern reproductive endocrinology emerged largely in Britain through the efforts of medical and agricultural scientists, with significant contributions by French, German, and later American investigators. Research began around gonadal extracts in 1889, although focus quickly shifted to nonreproductive hormones. This was due to the highly controversial nature of rejuvenescence claims made about the effects of gonadal extract therapy by Charles-Edouard Brown-Sequard, one of the "fathers" of reproductive sciences. Over thirty years passed before Edgar Allen and Edward Doisy first isolated a reproductive hormone.[27] In the interim, there were other notable developments: work by George Murray on remission of myxedema after doses of thyroid extract (1891); work by Edward Sharpey-Shafer and George Oliver on the vasorepressive effects of adrenal extract (1894); work by William Bayliss and Ernest Starling on pancreatic extract (1902); and the preparation of insulin by Frederick Banting and Charles Best (1922).[28] The disruption of British reproductive research by World War I, along with research of American investigators described below, led to enhanced American participation in the development of reproductive sciences.

The Emergence of Reproductive Sciences From Embryology

In sharp contrast to Britain, American reproductive research initiatives derived from embryological work in biology, medicine, and agriculture. The vast majority of the first generation of American reproductive scientists pursued embryological problems prior to and/or during their reproductive work, and several were led to reproductive problems via embryological investigations. Such investigators in biology included Frank Lillie, Herbert McLean Evans, George W. Bartelmez, Edward Everett Just, Carl G. Hartman, and Emil Witschi; in medicine, Philip Edward Smith, Charles Stockard, and George Papanicolaou, George W. Corner, and Edgar Allen; and in agriculture,

Leon J. Cole, Frederick F. McKenzie, and Clair E. Terrill. Even some latecomers to the field began with explicitly embryological work, including biologists Gregory Pincus and M. C. Chang.[29] Here I focus on two classic researches both of which were turning points in the development of American reproductive sciences: Lillie's work in the freemartin calf and Stockard and Papanicolaou's work on the estrus cycle of the guinea pig.

The Freemartin Research

Lillie's research on the freemartin was perhaps the most famous piece of reproductive science to come out of the Department of Zoology at the University of Chicago. The bovine freemartin was, according to Lillie, a "natural experiment." Lillie's research ultimately revealed that a freemartin is a sterile, female co-twin to a male, from a separate egg, but whose chorionic vessels (placentas) have merged in utero allowing crossing of blood systems.[30]

Research began in 1914 when the manager of Lillie's private farm sent him a pair of twin calf fetuses with their membranes.[31] Leon J. Cole, of the Department of Experimental Breeding in the College of Agriculture, University of Wisconsin, heard of Lillie's work on this problem and contacted him in hopes of correlating their efforts. Cole's department was working on multiple births in cattle: twins, double monsters, and freemartins.[32] There was considerable sharing of materials between the two centers because the freemartin problem was important to both, though from somewhat different standpoints.[33]

The first problem Lillie addressed was whether the twins came from the same or separate eggs. Prior research had argued for male twins from the same egg, largely because the twins were monochorial (attached to a single placenta), usually associated with one-egg twins. However, it did not make sense to Lillie that only one of a pair of male twins would be affected in utero. Lillie, who began to examine corpora lutea (sites of recently released eggs) in the ovaries, wrote Cole: "I am faced with the irritating difficulty that most of the uteri are received with one or both ovaries missing," making determination as to whether there had been one or two eggs released impossible.[34] Sufficient material gradually amassed to demonstrate that consistently there had been two corpora lutea (one in each ovary), and that there had been a fusion of two originally separate chorionic vessels (placentas) in utero. Since the freemartin usually possessed mammary glands and female as well as male external genitalia, and since if it were a male, sex ratios would be strangely skewed, Lillie concluded that it began as a female.[35]

In 1916 both Cole and Lillie published early abstracts in *Science*. Cole focused on sex ratios, reflecting his more genetic concerns.[36] Lillie focused on sex differentiation through the exchange of blood between fetuses: "If one is male and the other female, the reproductive system of the female is largely suppressed, and certain male organs even develop in the female. *This is unquestionably to be interpreted as a case of hormone action.* It is not yet determined whether the invariable result of sterilization of the female at the expense of the male is due to more precocious development of the male hormones, or to a certain natural dominance of male over female hormones."[37]

Lillie published his classic paper on the freemartin problem in 1917.[38] He emphasized that a vascular connection between the fetuses is requisite for development of a freemartin, and that influences of blood-borne hormones were acting on extant rudiments in the bisexual embryo stage. He concluded "that the course of embryonic sex-differentiation is largely determined by sex-hormones circulating in the blood."[39]

Such research continued at Chicago for some years.[40] Lillie noted that the work had "wider application than we expected . . . embryonic and astomoses blood vessels may have quite different results in different animals."[41] Lillie also found that intersexes in goats and swine may be genetic.[42] He had thought in 1917 that sex hormones were intensifiers of gene action, but by 1932 subsequent research had demonstrated the complete absence of sex differentiation in the absence of sex hormones.[43]

The freemartin work generated considerable interest in agricultural, popular science, and medical circles. One popular article noted: "Twins in cattle may be about two percent of all births in some breeds and the two sexed twins form about half of all the twin births, making the matter of sterile cows that produce no milk of economic importance to the dairy industry."[44] This work deepened connections among Lillie's group and agriculturally oriented centers of reproductive science headed by Cole at Wisconsin and F. A. E. Crew at Edinburgh.[45]

Lillie's work is also cited as pathbreaking in terms of developing theories of immunological tolerance important in medicine.[46] Such concerns were indicated by a popular science service's request to Lillie for a simple account of it in 1922: "There is a wide public interest just now in the subject of endocrinology. In fact the public seems . . . to take it up as a fad in succession to the Freudian complexes now going out of fashion."[47]

In sum, the importance of the freemartin work was multifold. First, it clearly demonstrated hormonal influence on sex differentiation in utero. Thus the production of sex, a classic turn-of-the-century biological problem, included not only genetic but also physiological processes. Second, the freemartin research "introduced biologists to the problems of the nature, origin, and action of sex hormones at a time when almost nothing was known about the subject."[48] One might even say that Lillie imported endocrinology into the embryology of his day. Third, problems of sexual differentiation that the work posed were central to several other major reproductive investigators including Emil Witschi, Carl R. Moore, and Dorothy Price.[49]

Problems of sex determination versus sex differentiation are classic examples of "boundary line" research. Susan Leigh Star and James R. Griesemer find that such problems often become "borderline sinkholes," unresolved and commonly ignored.[50] In fact, some years after his original work, Lillie found it necessary to reassert that "we must make a radical distinction between" the two, with sex determination as chromosomal (genetic) and sex differentiation, in higher animals, as hormonal.[51] More than fifty years later, John Farley, Dorothy Price, and Aubrey Gorbman point out that aspects of this problem have yet to be clarified.[52]

The Vaginal Smear

The second classic embryological investigation leading to reproductive research was that of Papanicolaou at Cornell Medical College in New York City.[53] Charles Stockard, a zoologist by training who worked in teratology while a student of Thomas Hunt Morgan,[54] was interested in the influence of chemicals on developing embryos. He began studying effects of alcohol on guinea pig embryos in 1909. Papanicolaou joined him in these efforts as an assistant in anatomy in 1914. Papanicolaou had previously worked on sex determination in Munich, and he resumed his studies using guinea pigs from the same colony supplying Stockard's research. While historical accounts are in some conflict, it seems most likely that Papanicolaou then initiated the use of vaginal examination and vaginal cell smears on slides in order to ascertain the estrus cycle stage to obtain ova at very precise stages of development.[55] This involved studying cells in the guinea pig vagina throughout the estrus cycle to determine whether stages of estrus could be indicated by the changing composition of those cells over the cycle.

The answer was a clear yes, but the implications of this research for reproductive investigations went far beyond the initially modest goals. The stages of the estrus cycle could be determined by microscopically examining smears of easily accessible cells. This could be done on a routine basis without surgery or sacrificing the animal. Moreover, such cells could serve as indicators of changes from normal phasing due to ablation (removal) of hypothesized hormone-producing organs, and/or transplantation of hormone-producing organs, and/or injection of hypothesized hormone extracts into the animal. And the indicator was not expensive to obtain, nor were the animals.

Each of these dimensions gave investigators increased latitude and flexibility. They could "infer what was happening in the internal reproductive organs without inspecting them directly," an extraordinarily powerful discovery.[56] Because so little was known about the sequence of events of the estrus and menstrual cycles, not until these events were determined and cataloged could experimenters gauge the alterations effected by either ablation or transplantation. Histologists painstakingly determined these gradual changes. The significance of the contributions of German and Austrian gynecologists lay precisely on this point. In the 1870s they cataloged the effects of female castration, during the period when oophorectomy was popular as a treatment for dysmenorrhea and certain neuroses. Later, Emil Knauer (in 1896) and Josef Halban (in 1900) noted that the ovary probably produced some special substance that normally maintained the uterus.[57] The vaginal smear technique permitted for the first time systematic examination of the biological activity of different organs imputed to produce hormones.

Ironically, having such accurate, cheap, and accessible indicators in laboratory animals initially was viewed as important to the production of laboratory animals in colonies, especially but not exclusively for embryological research![58] Implications of the technique were appreciated quite quickly, and it became central to many reproductive endocrinological investigations. For example, while developing a colony of mice for a course in embryology, some of the implications of his morphological work drew Edgar Allen into pioneering investigations in reproductive endocrinology. He and Edward Doisy, a biochemist, soon achieved the first isolation of a reproductive hormone, active ovarian follicular extract.[59]

The smear method also became a primary means of studying estrus and menstrual cycles in different animals. Here Stockard and Papanicolaou's work was followed quickly by Long's and Evans's parallel efforts on the rat, Edgar Allen's on the mouse, Corner's on the monkey, sow, and swine, Hartman's on the opossum, and Evans's and Cole's work on the dog. Agricultural animals studied included the cow, ewe, mare, and sow.[60] As Evans later noted, "It appeared, indeed, for a time that the application of the vaginal smear method would be all that was required to segment the stages of the estrous cycle in all animals . . . [but] the beautifully distinct changes in the vaginal lochia of small rodents were peculiar for the smaller forms."[61] Only in the dog was the estrogen level high enough for pronounced vaginal changes. But Papanicolaou (1933) pursued the vaginal smear as a potential indicator of something in women. He ultimately found that what later became known as the "Pap Smear" could indicate potential and actual pathological changes in the cervix and uterus useful for diagnosis of cancerous, precancerous, and other abnormal conditions.[62]

In addition to the two classic investigations of Lillie and of Stockard and Papanicolaou, the importance of embryological work for the development of reproductive sciences was reflected in the inclusion of a major chapter on embryology in *Sex and Internal Secretions,* the American Bible of reproductive sciences first published in 1932.[63] George Corner, in his foreword to the third edition in 1961, stated: "To the embryologists of Europe and America we owe in large part also the successful analysis of the mammalian reproductive cycle that has been achieved during this half century."[64]

It is surprising that problems of reproductive physiology and endocrinology were predominantly pursued in the United States in the early twentieth century not by physiologists but by zoologists and anatomists, most with backgrounds in embryology. Addressing this historical problem, Diana Long undertook a statistical analysis of the disciplinary affiliations and identities of "sex researchers" publishing in the *American Journal of Physiology* from 1923 to 1947. Although only about one-third identified themselves primarily as physiologists, two-thirds were members of the American Physiological Society. Long argues that sex researchers, regardless of discipline, gained many benefits through associating with the prestigious physiological society and journal.[65] My own work indicates that reproductive endocrinology was linked both literally and in terms of enhanced legitimacy to the more prestigious field of general endocrinology through shared study of the anterior pituitary gland which produced both reproductive and nonreproductive hormones. Prestige and legitimacy were very important assets for early reproductive scientists whether they considered themselves to be primarily anatomists, zoologists, physiologists, or endocrinologists.[66]

In sum, the freemartin work supported the endocrinological direction of American reproductive sciences. Then, because of its value as an indicator, the vaginal smear work led many investigators from different disciplines and professions into a new domain of research. By 1925, reproductive sciences were fully initiated as integrated lines of modern scientific work, with emphasis in the field shifting from classical physiological work to biochemical endocrinology. The period from 1925 to 1940 was characterized by British researcher A. S. Parkes as the "endocrinological gold rush," and by Guy Marrian as the "heroic age of reproductive endocrinology."[67]

Doability and the Rise of Reproductive Sciences

By World War I, the emergent reproductive sciences were viewed as controversial by some constituencies owing to their association with sexuality, clinical quackery, social movements such as birth control advocacy, and because of their capacity to alter "natural" patterns of reproduction. Why would so many investigators of considerable stature and renown pursue reproductive research between the wars despite its controversial status? I argue here that they did so because reproductive research was, after 1917, highly "doable" research especially when compared with that in embryology at the time.

Joan H. Fujimura recently put forward the concept of the "doability" of research.[68] This refers to whether or not a specific line of work (a set of investigations focused on a particular set of problems) is feasible and worthwhile to undertake at a specific time and place. To construct doable problems, investigators must assess the particular needs for doing the actual experiments or other aspects of investigation, the work organization and commitments of the laboratory or other research site, and the available support of various kinds for doing that work in wider scientific and related worlds. Doability thus requires investigators to align or fit their research problems simultaneously across experimental capacities, laboratory organization and direction, and broader social worlds of fiscal, scientific, and extrascientific support. Before beginning the work, scientists must both pull together and articulate a wide array of requisite elements to make as sure as possible, given the circumstances, that something recognized as worthwhile will emerge downstream.

The concept of doability provides a framework for examining the rise of reproductive sciences after the three-way split. The question is whether the emergent and newly bounded disciplines (genetics, developmental embryology, and reproductive sciences) offered researchers equally attractive lines of highly "doable" research after 1910. For each field, the question must be answered at the levels of the experiment, laboratory, and wider social worlds.

Doability in Genetics

Doable research in genetics preceded that in reproductive physiology. The key to increased genetics research at the experiment level after 1910 was the use of *Drosophila,* corn, guinea pigs, and chickens as major research materials, the right tools for the job.[69] As Ross Harrison commented, "Much progress has depended upon the fortunate findings of organisms that illustrate this or that principle clearly or such as submit to the most ruthless experimentation."[70]

In terms of research personnel at the laboratory level, histories of genetics fully document the adequacy of staffing.[71] What Fujimura calls the social world level at which doability must also be assessed is more complex. It includes the likelihood of research payoffs from pursuit of the line of work, the wider scientific legitimacy of pursuit of those problems, along with fiscal and other kinds of support for research. It also includes, with varying degrees of salience, the social legitimacy of those problems. In genetics, the initial investigations that triggered the three-way split were immediate proof of the doability and probable high payoffs of genetics research. The scientific legitimacy of problems of heredity and evolution had been well established for centuries and intensified after Darwin.[72] Genetics research was quite cheap, so fiscal support was not a serious issue. Wider scientific support was extensive among both agricultural breeders and biologists. Through enhanced control over hereditary processes in plants and animals, genetics promised tremendous profitability downstream. Moreover, the eugenics movement provided broad social legitimacy for enhanced control over heredity in humans. By 1915, T. H. Morgan and his associates' *The Mechanism of Mendelian Heredity* fully established this line of work.[73]

Doability in Reproductive Sciences

Highly "doable" research on reproduction began with Stockard and Papanicolaou's work in 1917. The vaginal smear technique for obtaining an indicator of estrus stage was as potent for the development of reproductive sciences as the visible chromosome structure of *Drosophila* was for genetics. Taking speedy advantage of applicable results such as the smear technique is an important aspect of constructing doable problems. The freemartin investigations of Lillie and his associates immediately linked this work to cutting-edge endocrinological problems. Both studies triggered the flood of reproductive investigations in biology, medicine, and agriculture noted above. The development by Philip Edward Smith after 1916 of hypophysectomy techniques (to remove surgically the anterior pituitary without brain or other damage) permitted finer-grained assessments of the biological activities of hormones, often in conjunction with the use of vaginal smears.[74]

In the laboratory, the biochemical nature of the endocrinological thrust of reproductive sciences quickly led investigators into collaboration with biochemists.[75] At different research centers, these collaborations took different forms, although the reproductive scientists rather than the biochemists seem to have established the research agendas.[76] For example, George Corner at Rochester hired biochemist Willard Allen. Similarly, H. L. Fevold served as a research chemist in the laboratory of Frederick Hisaw, first at Wisconsin and later at Harvard. In Herbert Evans's institute at Berkeley the chemist was Choh Hao Li. Lillie and his colleagues worked closely with Fred C. Koch of the Department of Physiological Chemistry. And Edgar Allen at Washington University in St. Louis worked closely with Edward A. Doisy, professor of Biochemistry in the School of Medicine.[77]

However, the greatest change occurred at the wider social world level, in terms of scientific and social legitimacy and support. Beginning in the 1910s and increasing after World War I, the social movements of birth control, eugenics, and neomalthusianism provided organized markets for reproductive sciences, serving as audiences, sponsors, and consumers of research. Moreover, by raising reproductive topics as appropriate to public forums, prestigious activists from these movements, including many scientists, also countered the illegitimacy that had slowed the development of reproductive research.[78] Scientific study of human populations also began to establish the propriety of such topics in the academy and society. What Merriley Borell has noted in the case of Britain was also true in the United States: the association of reproductive endocrinology "with human sex and reproduction affected the nature of the search itself, causing important fluctuations in the progress of research in this area."[79]

The upsurge of fiscal and social support for reproductive research during the interwar years came through the National Research Council's (NRC) new Committee for Research on Problems of Sex, which existed from 1921 to 1962.[80] From 1921 to 1940, the committee sponsored reproductive research through grants totaling $1,087,322 funded almost exclusively by Rockefeller monies.[81] This committee essentially paid for the American development of reproductive endocrinology.[82]

The committee was initially charged with studying human sexuality, but Lillie and other members of the committee of a more basic biological research persuasion successfully redirected its mission and funding into biological research on reproduction for its first twenty years. By capturing the committee, these scientists simultaneously gained broad legitimation for reproductive sciences through association with the NRC, significant funding, direct relationships between reproductive scientists and a major biomedical research sponsor, and a strong "basic" research identity.[83]

Rather than providing grants to individuals at many different institutions, the committee operated by funding established and emerging centers of reproductive research staffed by multiple researchers under the leadership of an investigator of considerable stature. Many of the major investigators of the early years (including Frank Lillie, Charles Stockard, Herbert Evans, Emil Witschi, and Edgar Allen) had done considerable embryological work. New investigators who focused more exclusively on endocrinology included Philip E. Smith, Frederick Hisaw, and Edward Koch. By the 1930s, research agendas had expanded to include sex behavior, sex psychology (often psychobiology), and related neurological problems in animals and humans under the direction of Robert M. Yerkes, Lewis M. Terman, Philip Bard, Frank A. Beach, Walter B. Cannon, Gladwyn Kingsley Noble, and Calvin P. Stone. During the 1930s, the committee also sponsored Gregory Pincus's work on in vitro fertilization and an array of researches that eventually led to the development of the contraceptive pill.[84] Reproductive sciences were quite interdisciplinary, including a wide array of researchers under a flexible umbrella.

In 1934 Warren Weaver vividly explained the Rockefeller Foundation's interest in the work of the Committee for Research on Problems of Sex: "Can man gain an intelligent control of his own power? Can we develop so sound and extensive a genetics that we can hope to breed, in the future, superior men? Can we obtain enough knowledge of physiology and psychobiology of sex so that men can bring this pervasive, highly important and dangerous aspect of life under rational control?"[85] Reproductive scientists answered loudly in the affirmative.

Thus powerful sectors in American society sought and supported enhanced scientific control over nature in many forms.[86] Reproductive sciences promised enhanced control over reproduction in animals and humans alike. Control over reproduction was, of course, deeply linked to issues of enhancing control over life more generally, a theme appropriately taken up by Allen, Beatty, Paul, and Burkhardt, and Mitman in this volume. Pauly has argued that "controlling life" was at the heart of Jacques Loeb's research, including but far from limited to control over reproduction.[87] Such issues of control are tied to economic and social control processes as many kinds of reformers adapted the means and mechanisms of industrialization first developed in factories and sought to fit them not only into agricultural and social life but also into the sciences themselves early this century.

More specifically, while Loeb articulated a scientific philosophy of control, by the 1930s such Loebians as Pincus, Herman J. Muller, and B. F. Skinner were successfully articulating and disseminating such a doctrine. This philosophy placed science primarily in the service of the economy while permitting scientists to retain some shreds of the autonomy so valued by Loeb himself. The "climate" that nurtured the work of Pincus was thus partly Loebian but also much more vaunting of the scientific promise of economic benefits and social control. It is in this broader context of expanding the production of scientific research and its downstream applications that doability must also be assessed. Like industrialization, doability requires enhanced control over the processes of production.

Doability in Embryology

But doability is a relative phenomenon and must be assessed comparatively. This was especially the case at the turn of the century due to the highly porous nature of disciplinary and professional boundaries that permitted investigators to move easily in new directions. The intriguing question here is whether investigations in genetics and reproductive sciences were particularly attractive as alternatives to those under way and envisioned in embryology at the time and, if so, why. This in turn raises the broader question of what was going on in embryology after 1910.

It can certainly be argued that, after the flood of experimental work in 1890–1910, embryology experienced a slump between 1910 and 1925. Ross Harrison of Yale, a leading American embryologist, asserted in 1937 that embryology had been doing poorly from 1910 to 1925, the very years of the emergence of American reproductive sciences and the dramatic growth in genetics: "The fertility of the soil [in embryology] seemed to have suddenly run out and tillage no longer worth while. What more human, then, than the gold rush to genetics and general physiology. . . . Later came another gold rush to endocrinology, now perhaps at its height."[88] Benjamin H. Willier's and Jane M. Oppenheimer's collection of papers also illustrates the slump in embryology. Their volume includes six articles from 1888 to 1908, nothing between 1908 and 1913, only four articles between 1913 and 1924, and nothing again until 1939. Moreover, two of the articles between 1913 and 1924 were by Lillie, one on fertilization and another on the freemartin. Both can be viewed as works in the physiology of reproduction as much as, if not more than, in embryology.[89] During this period, doable lines of work in embryology failed to produce advances, and the field was viewed as comparatively dull: "embryology . . . appeared increasingly to be in some disarray."[90]

While embryology was attracting fewer scientists, both genetics and reproductive sciences began to attract more, and younger and more established scientists were drawn into these emergent lines of work. For example, Garland Allen stresses that "the agricultural climate around the turn of the century . . . emphasized genetic transmission rather than embryonic differentiation as the crucial problem to be understood."[91] Plant genetics provided agricultural payoffs, furthering its appeal to scientists. Control over genetic processes in animal production—the focus of animal agriculture—requires a well-developed understanding of reproductive phenomena. Some control over male genetic contributions was achieved through artificial insemination, and agricultural interests supported development of reproductive sciences.[92]

Like genetics, reproductive sciences were an "open territory" drawing investigators into a new field as American biology expanded. Investigators who went into either could construct doable problems. They could meet immediate research needs, produce good scientific work with relative ease, obtain funding and other kinds of legitimacy, develop their individual careers, build laboratories as centers of focused research, and help found new disciplines and subdisciplines. Clearly both genetics and reproductive sciences profited greatly from embryology's loss. The ability to pull the requisite pieces together to construct doable problems at a given historical moment can be highly consequential. We can thus see how the concrete capacity to construct doable problems contributes to patterns of scientific change.

Conclusions: The Three-Way Split Revisited

During an era when there were relatively few researchers in the life sciences,[93] a significant number of highly capable investigators pioneered the development of reproductive sciences. By 1940, with the second generation of investigators, they had established a number of major research centers, garnered extensive funding from a prestigious NRC committee, and achieved preeminence for the United States in yet another scientific domain.

Why, then, have reproductive sciences been comparatively ignored in histories of biology, medicine, and even agriculture? Why too have studies of the splitting of embryology into emergent disciplines attended only to genetics and developmental embryology, omitting consideration of reproductive sciences? There are several reasons.

First, the illegitimacy of pursuing reproductive sciences itself carries over into history and sociology of science.[94] Beyond invaluable accounts of their work by reproductive scientists, remarkably little has been done in this area despite global impacts of developments such as the Pill and IUD, impacts at least equal to those of genetics or developmental embryology. Maienschein notes the lack of attention to botany in turn-of-the-century biology and the parallel relative absence of histories of botany. She attributes this to botany not being a "sexy" enough area of research. [95] If botany is not "sexy" enough, reproduction is too sexy, literally as well as metaphorically.

Second, since World War II, reproductive sciences have been pursued predominantly in medical and agricultural institutions and contexts.[96] Historians and laypeople alike conceive reproductive sciences as primarily medical and secondarily agricultural research and practice areas. Therefore the history of reproductive sciences has often been defined as being part of the history of medicine or agriculture, both of which are underdeveloped areas of twentieth-century studies. Hence the presence of lively and eminent centers of reproductive sciences in major biological departments and schools across the country in the early decades of this century, and the centrality of biologists to American reproductive sciences before World War II, have also been rendered invisible.

Third, biologists have also ignored reproductive sciences and their implications, thereby shaping historians' views of "what counts" to some degree. For example, Evelyn Fox Keller has recently has recently argued that the failure of evolutionary theory to take reproductive processes into account has serious consequences for the robustness of evolutionary theory. Elisabeth Lloyd similarly found that despite careful citations to studies of human sexuality in evolutionary theoretical treatises, the actual contents of those studies were ignored or distorted.[97]

But ultimately, the problem is not merely that reproductive sciences and the three-way split from embryology have been relatively ignored. Rather, as historians and historically minded philosophers, sociologists, and others move into studies of sciences in the twentieth century, we need to rethink aspects of the historical project to fit the ways in which scientists actually worked. We need both disciplinary and interdisciplinary histories, focused on high- and low-status scientific research areas. We must grasp the full range of variation of life sciences work, core and periphery, minor as well as major research traditions. For both disciplinary and professional boundaries were so porous at the turn of the century that waves of new Ph.D.s from biological disciplines ended up in medical and agricultural as well as biological settings. We must follow them and examine their research and institutional situations without *a priori* assumptions about distinctions between "basic" and "applied" research. For, I would argue, almost all research done in twentieth-century America has been pursued or at least justified on practical grounds regardless of where it was pursued. The practicality of the split between developmental embryology and genetics[98] was not merely a matter of efficient research organization or competition for resources but also a reorganization that clarified the productive capacity of a new line or work—genetics—which would prove itself in terms of profitability. Nor did reproductive sciences hide behind the skirts (or more accurately the trousers) of embryology but, despite moral opprobrium, initially claimed identity boldly as "the biology of sex" for similar practical payoffs.[99]

"What counts" in American life sciences may itself be distinctive. As Horder and his associates have noted, "All the more intriguing then is the question of why it was in the United States that the 'split' between embryology and genetics occurred so prominently."[100] The splitting off of reproductive sciences was also prominent in the United States. Indeed, if we seek an "ecology of knowledge"[101] and the conditions of its production, we must also ask "What counts to whom and under what conditions?" Ultimately, comparative work will help answer such questions about the nature of scientific change and the patterns of movement at the "cutting edge" in twentieth-century life sciences.

Acknowledgments

Special thanks for collegial support go to Howard Becker, Merriley Borell, Joan Fujimura, Diana Long, Gregg Mitman, Jane Maienschein, M. C. Shelesnyak, S. Leigh Star, Anselm Strauss, and to all the contributors to this volume whose comments on an earlier version were invaluable. This work was supported by an NIMH postdoctoral fellowship in the Department of Sociology, Stanford University and draws on earlier research supported by fellowships from the University of California, San Francisco and the Rockefeller University.

Notes

1. F. H. A. Marshall, *The Physiology of Reproduction* (London: Longmans, Green, 1910).

2. A basic summary of the development of American reproductive sciences appears in Adele E. Clarke, "A Social Worlds Research Adventure: The Case of Reproductive Science," in Susan Cozzens and Thomas Gieryn, eds., *Theories of Science in Society* (Bloomington: University of Indiana Press, 1990), pp. 23–50; see

also idem, "Emergence of the Reproductive Research Enterprise: A Sociology of Biological, Medical and Agricultural Science in the United States, 1910–1940" (Ph.D. dissertation, University of California, San Francisco, 1985); and idem, "Controversy and the Development of Reproductive Sciences," *Social Problems,* 1990, *37:* 18–37.

3. Garland E. Allen, *Life Sciences in the Twentieth Century* (New York: Cambridge University Press, 1978); and William Coleman, *Biology in the Nineteenth Century: Problems of Form, Function and Transformation* (New York: Cambridge University Press, 1977).

4. Both Gilbert and Allen have made this split a focus of their work. Scott F. Gilbert, "Cellular Politics: Ernest Everett Just, Richard B. Goldschmidt, and the Attempt to Reconcile Embryology and Genetics," in Ronald Rainger, Keith R. Benson, and Jane Maienschein, eds., *The American Development of Biology* (Philadelphia: University of Pennsylvania Press, 1988), pp. 311–346; idem, "In Friendly Disagreement: Wilson, Morgan, and the Embryological Origins of the Gene Theory," *American Zoologist,* 1987, *27:* 797–806; and idem, "The Embryological Origins of the Gene Theory," *J. His. Biol.*, 1978, *11:* 307–351. See also Garland E. Allen, "T. H. Morgan and the Split between Embryology and Genetics, 1910–1935," in T. J. Horder, J. A. Witkowski, and C. C. Wylie, eds., *A History of Embryology: The Eighth Symposium of the British Society for Developmental Biology* (Cambridge: Cambridge University, 1986), pp. 113–146.

5. Merriley Borell, "Biologists and the Promotion of Birth Control Research, 1918–1938," *J. Hist. Biol.*, 1987, *20:* 57–87; and idem, "Organotherapy and the Emergence of Reproductive Endocrinology," *J. Hist. Biol.*, 1985, *18:* 1–30. F. H. A. Marshall was well acquainted with the work of Walter Heape on reproductive phenomena; Heape had been a student of Balfour, the embryologist. Like Balfour's other students, he took his work in new and more causal directions. See Mark Ridley, "Embryology and Classical Zoology in Great Britain," in Horder, Witkowski, and Wylie, *A History of Embryology,* pp. 35–71, esp. p. 46.

6. On Germany, see George W. Corner, *Seven Ages of a Medical Scientist: An Autobiography* (Philadelphia: University of Pennsylvania Press, 1981). On France, see L. Brouha, ed., *Les hormones sexuelles: Colloque International, Conferences du College de France, Fondation Singer-Polignac* (Paris: Hermann, 1938); and David Hamilton, *The Monkey Gland Affair* (London: Chatto and Windus, 1986).

7. See, for example, Rainger, Benson, and Maienschein, *The American Development of Biology;* and Jeffery Werdinger, "Embryology at Woods Hole: The Emergence of a New American Biology" (Ph.D. dissertation, Indiana University, 1980).

8. Merriley Borell, "Setting the Standards for a New Science: Edward Schafer and Endocrinology," *Medical History,* 1978, *22:* 282–290; idem, "Organotherapy, British Physiology and Discovery of the Internal Secretions," *J. Hist. Biol.*, 1976, *9:* 235–268; and idem "Brown-Sequard's Organotherapy and Its Appearance in America at the End of the Nineteenth Century," *Bulletin of the History of Medicine,* 1976, *50:* 309–320.

9. Sophie D. Aberle and George W. Corner, *Twenty-Five Years of Sex Research: History of the National Research Council Committee for Research in Problems of Sex, 1922–1947* (Philadelphia: Sauders, 1953); and Clarke, "Controversy."

10. For example, Jane Maienschein, Ronald Rainger, and Keith R. Benson, "Introduction: Were American Morphologists in Revolt?" *J. Hist. Biol.*, 1981, *14*(1): 83–87; Garland E. Allen, "Naturalists and Experimentalists: The Genotype and the Phenotype," *Studies in the History of Biology*, 1979, *3:* 179–209; Jan Sapp, "The Struggle for Authority in the Field of Heredity, 1900–1932: New Perspectives on the Rise of Genetics," *J. Hist. Biol.*, 1983, *16*(3): 311–342; Charles Rosenberg, "Factors in the Development of Genetics in the United States: Some Suggestions," *Journal of the History of Medicine and Allied Sciences,* 1967, *22:* 27–46.

11. Frederick B. Churchill, "Sex and the Single Organism: Biological Theories of Sexuality in Mid-Nineteenth Century," *Studies in the History of Biology,* 1979, *3:* 139–178, on p. 140.

12. Sapp, "The Struggle for Authority."

13. Oppenheimer notes that certain geneticists lobbied against such restrictiveness, notably E. B. Wilson and Sewall Wright, and that Wilson's work led to cytogenetics and cytochemistry which subsequently merged into molecular biological approaches to similar problems. See Jane M. Oppenheimer, *Essays in the History of Embryology and Biology* (Cambridge, Mass.: MIT Press, 1967), pp. 35–38.

14. Churchill, "Sex and the Single Organism," p. 140. Many of these "geneticists" were themselves embryologically trained. See Garland E. Allen, "The Transformation of a Science: T. H. Morgan and the Emergence of a New American Biology," in Alexandra Oleson and John Voss, eds., *The Organization of Knowledge in Modern America, 1860–1920* (Baltimore: Johns Hopkins University Press, 1979).

15. Viktor Hamburger, "Embryology and the Modern Synthesis in Evolutionary Theory," in Ernst Mayr and William B. Provine, eds., *The Evolutionary Synthesis: Perspectives on the Unification of Biology* (Cambridge, Mass.: Harvard University Press, 1980), pp. 97–111.

16. Jane Maienschein, "What Determines Sex? A Study of Converging Approaches, 1880–1916," *Isis*, 1984, *75*(278): 457–480, on p. 480.

17. Marshall, *The Physiology of Reproduction*, p. 2.

18. His scope for the physiology of reproduction is examined in detail by Clarke, "Emergence," pp. 115–125.

19. Frank Rattray Lillie, *Problems of Fertilization* (Chicago: University of Chicago Press, 1919), p. 129, emphasis added.

20. Frank R. Lillie, "The History of the Fertilization Problem," *Science*, 1916, *42:* 39–53.

21. This was published as Appendix 2 in Aberle and Corner, *Twenty-Five Years of Sex Research*, pp. 102–104.

22. See, for example, Jane Maienschein, "History of Biology," *Osiris 2nd ser. 1:* 147–162, esp. p. 161.

23. Philip J. Pauly, "The Appearance of Academic Biology in Late Nineteenth Century America," *J. Hist. Biol.*, 1984, *17*(3): 369–397, esp. pp. 392–395; idem, "Summer Resort and Scientific Discipline: Woods Hole and the Structure of American Biology, 1882–1925," in Rainger, Benson, and Maienschein, *The American Development of Biology*, pp. 121–150.

24. Borell, "Setting the Standards," p. 286.

25. See, for example, Borell, "Organotherapy and the Emergence," p. 11.

Also Robert E. Kohler, *From Medical Chemistry to Biochemistry: The Making of a Biomedical Discipline* (Cambridge: Cambridge University Press, 1982). In Britain, the term "hormone" was used for chemicals with demonstrable physiological effects and was more in the domain of laboratory scientists; the term "internal secretions" was used to suggest a hypothetical entity whose absence resulted in disease and was more in the domain of clinicians. American usage of these terms does not seem so clear-cut.

26. Merriley Borell, "Origins of the Hormone Concept: Internal Secretions and Physiological Research, 1889–1905" (Ph.D. dissertation, Yale University, 1976). p. xii. These identity problems are quite complex; Clarke, "Emergence," 169–213.

27. Edgar Allen and Edward Doisy, "An Ovarian Hormone: Preliminary Report on Its Localization, Extraction and Partial Purification, and Action," *Journal of the American Medical Association,* 1923, *81:* 819–821.

28. Borell, "Organotherapy and the Emergence."

29. See Clarke, "Emergence," pp. 131–147.

30. Frank Rattray Lillie, "The Free-Martin; A Study of the Action of Sex Hormones in the Foetal Life of Cattle," *Journal of Experimental Zoology,* 1917, *23:* 371; idem, "Sex-Determination and Sex-Differentiation in Mammals," *Proceedings of National Academy of Science,* 1917, *3:* 464–470. See also B. H. Willier, "Frank Rattray Lillie, 1870–1947," *Biographical Memoirs of the National Academy of Science, U.S.A.*, vol. 30 (New York: Columbia University Press, 1957), pp. 179–236.

31. Newman dates this as 1916 but was likely mistaken; see H. H. Newman, "History of the Department of Zoology in the University of Chicago," *Bios*, 1948, *19*(4): 215–239, esp. p. 231. Willier dates the onset of Lillie's freemartin work as autumn 1914; see Willier, "Lillie," p. 216. And Leon J. Cole, who specialized in experimental breeding at Wisconsin, was already in correspondence with Lillie on freemartins in late 1915 (Cole to Lillie, 17 November 1915, and passim, Frank Rattray Lillie Papers [hereafter Lillie Papers], Box 2, Folder 15, Archives, University of Chicago). This was, of course, not the first research on freemartins, since the sterility of such a cow was important to farmers. For a description of earlier work, see Thomas R. Forbes, "The Origin of the Freemartin," *Bulletin of the History of Medicine*, 1946, *20:* 461–466.

32. He had a collection of postembryonic freemartin gonadal material for study, complementing Lillie's embryos and fetuses. Lillie responded that cooperation would be valuable, and that he had already amassed twenty-five pairs of embryonic twins. Lillie to Cole, 22 November 1915, Lillie Papers, Box 2, Folder 15.

33. Cole to Lillie, 21 February 1917, Lillie Papers, Box 2, Folder 15. Lillie and his group relied primarily on the Chicago stockyards for freemartin material, acquired through a particular foreman at a Swift and Co. abattoir and through the special efforts of the department's collector (Willier, "Lillie," p. 215). Lillie noted: "every uterus containing twins below a certain size from a certain slaughter house is sent to me for examination without being opened" (F. R. Lillie, "The Theory of the Free-Martin," *Science,* 1916, *43:* 611–613, esp. 612). However, this was not always smooth. In 1918 Lillie noted: "I am having no success in securing new material from the stockyards. New federal regulations and reduction in the amount of business done there combined to make the embryonic material very scarce" (Lillie to Cole, 13 January 1918, Lillie Papers, Box 2, Folder 15).

34. Lillie to Cole, 3 December 1915, Lillie Papers, Box 2, Folder 15.

35. Lillie, "The Free-Martin," and Lillie, "Sex-Determination."

36. Leon J. Cole, "Twinning in Cattle with Special Reference to the Freemartin," *Science*, 1916, *43:* 177.

37. Lillie, "The Theory of the Free-Martin," p. 612, emphasis added. See also Nelly Oudshoorn, "Endocrinologists and the Conceptualization of Sex, 1920–1940," *J. Hist. Biol.*, 1990, *23:* 163–186.

38. Lillie, "The Free-Martin."

39. Ibid., p. 415. Lillie's conclusions have subsequently been modified. See Willier, "Lillie," p. 219; Benjamin H. Willier and Jane M. Oppenheimer, eds., *Foundations of Experimental Embryology* (Englewood Cliffs, N.J.: Prentice-Hall, 1964), pp. 136–143, esp. pp. 137–138; and Dorothy Price, "Mammalian Conception, Sex Differentiation, and Hermaphroditism as Viewed in Historical Perspective," *American Zoologist,* 1972, *12:* 179–191.

40. The focus was upon collecting specimens ranging in size/age for comparative purposes (including adult freemartins), and on seeking the range of anatomical and physiological variation in freemartins (including fertility). T. H. Bissonnette to Lillie, 4 August 1922, Lillie Papers, Box 2 Folder 3.

41. Lillie to Bissonette, 21 May 1928, Lillie Papers, Box 2, Folder 3.

42. Lillie to Dr. C. J. Elmore, 26 October 1928, Lillie Papers, Box 3, Folder 6.

43. Lillie, "General Biological Introduction," in Edgar Allen, ed., *Sex and Internal Secretions: A Survey of Recent Research* (Baltimore: Williams and Wilkins, 1932), pp. 1–11, esp. p. 6. For a discussion of relations between genetic and endocrine factors in reproduction, see Charles H. Danforth, "Interrelation of Genetic and Endocrine Factors in Sex," in Allen, *Sex and Internal Secretions,* pp. 12–54.

44. "The Mystery of Sterile Twin Heifer Solved," by Science Service, hand-dated 17 July 1922, Lillie Papers, Box 6 Folder 3.

45. Crew to Lillie, 4 April 1928, Lillie Papers, Box 2, Folder 20. And Bissonnette to Lillie, 1 January 1929, Lillie Papers, Box 2, Folder 3. On Crew and his work, see Lancelot Hogben, "Francis Albert Eley Crew, 1886–1973," *Biographical Memoirs of Fellows of the Royal Society,* 1974, *20:* 134–153.

46. Rupert E. Billingham and Alan E. Beer, "Reproductive Immunology: Past, Present and Future," *Perspectives in Biology and Medicine*, 1984, *27:* 259–275.

47. Edwin E. Slosson to Lillie, 28 June 1922, Lillie Papers, Box 6, Folder 3.

48. Willier, "Lillie," p. 219.

49. Aubrey Gorbman, "Emil Witschi and the Problem of Vertebrate Sexual Differentiation," *American Zoologist,* 1979, *19:* 1261–1270; Carl Moore, "Biology of the Testis," in Allen, *Sex and Internal Secretions,* pp. 281–371; Price, "Mammalian Conception."

50. Susan Leigh Star and James R. Griesemer, "Institutional Ecology, 'Translations' and Boundary Objects: Amateurs and Professionals in Berkeley's Museum of Vertebrate Zoology, 1907–1939," *Social Studies of Science,* 1989, *19:* 387–420; and Maienschein, "History of Biology," p. 161.

51. See Lillie's "Biological Introduction," p. 5.

52. John Farley, *Gametes and Spores: Ideas about Sexual Reproduction, 1750–1914* (Baltimore: the Johns Hopkins University Press, 1982), pp. 259-263; Price, "Mammalian Conception"; and Gorbman, "Witschi."

53. Charles R. Stockard and George Papanicolaou, "A Rhythmical 'Heat Period'

in the Guinea-Pig." *Science,* 1917, *46:* 42–44; Charles R. Stockard and George Papanicolaou, "The Existence of a Typical Oestrous Cycle in the Guinea-Pig, with a Study of Its Histological and Physiological Changes," *American Journal of Anatomy,* 1917, *22:* 225–283. See also John F. Fulton and Leonard G. Wilson, *Selected Readings in the History of Physiology,* 2nd ed. (Springfield: Charles C. Thomas, 1966), pp. 395–398.

54. Jane M. Oppenheimer, "Basic Embryology and Clinical Medicine: A Case History in Serendipity," *Bulletin of the History of Medicine,* 1984, *58:* 236–240.

55. The historiographic conflict is between the accounts of Oppenheimer and Carmichael. Oppenheimer, like more general sources, attributes the work to both Stockard and Papanicolaou, as published. She further argues that the smear was undertaken in relation to improving "controls" for the alcohol studies. In sharp contrast, Carmichael quotes Papanicolaou as saying it was part of his own sex determination research but was published under both names because that was the custom. Papanicolaou became distraught when Stockard was given the bulk of the credit by colleagues at Woods Hole, and raised the matter with Stockard who said that his name need no longer be added to Papanicolaou's own work. See Oppenheimer, "Basic Embryology," p. 238; idem, "Some Historical Relationships between Teratology and Experimental Embryology," *Bulletin of the History of Medicine,* 1968, *42:* 145–159; D. Erskine Carmichael, *The Pap Smear: Life of George N. Papanicolaou* (Springfield: Charles C. Thomas, 1973), esp. pp. 44–53.

56. Sir Solly Zuckerman, *Beyond the Ivory Tower: The Frontiers of Public and Private Science* (New York: Taplinger, 1970), p. 22.

57. Borell, "Organotherapy and the Emergence," p. 12; and George W. Corner, "The Early History of the Estrogenic Hormones," *Endocrinology* (Proc. Soc.), 1965, *31:* iii–xvii.

58. Colonies were not very well established, and improvements were constantly being sought. The smear permitted prompt and effective mating. See William C. Young, ed., *Sex and Internal Secretions,* 3rd ed., vols. 1-2 (Baltimore: Williams and Wilkins, 1961), p. xiv; and Adele E. Clarke, "Research Materials and Reproductive Science in the United States, 1910–1940," in Gerald L. Geison, ed., *Physiology in the American Context, 1850–1940* (Bethesda: American Physiological Society, 1987), pp. 323–350.

59. Allen and Doisy, "An Ovarian Hormone"; and E. A. Doisy, "Biochemistry of the Follicular Hormone Theelin," in Allen, *Sex and Internal Secretions,* pp. 481–498.

60. The classic papers are J. A. Long and H. M. Evans, "The Oestrus Cycle in the Rat," *Anatomical Record,* 1920, *18:* 241–248; Edgar Allen, "The Oestrus Cycle in the Mouse," *American Journal of Anatomy,* 1922, *30:* 297–372; and George W. Corner, "Cyclic Changes in the Ovaries and Uterus of the Sow, and Their Relation to the Mechanism of Implantation," *Carnegie Contributions to Embryology,* 1921, *13*(64): 117–146. See also Clarke, "Emergence."

61. Herbert McLean Evans, "Foreword," H. H. Cole and P. T. Cupps, *Reproduction in Domestic Animals* (New York: Academic Press, 1959), p. vii.

62. George N. Papanicolaou, "The Sexual Cycle in the Human Female as Revealed by Vaginal Smear," *American Journal of Anatomy,* 1933, *52:* 519–637. He also attempted an early pregnancy diagnosis through smears, but the Ascheim-Zondek urine-based test proved more successful. See Carmichael, *The Pap Smear.*

63. Allen, *Sex and Internal Secretions,* pp. 94–159. Because work was interrupted during both World Wars, there were problems in writing and publishing later editions of Marshall's (British) text, leaving the field more open to the Americans. See F. H. A. Marshall, *The Physiology of Reproduction,* 2nd ed. (London: Longmans, Green, 1922); and A. S. Parkes, *Marshall's Physiology of Reproduction,* 3rd ed. (New York: Longmans, Green, 1952–1966).

64. George W. Corner, "Foreword," in Young, *Sex and Internal Secretions,* pp. ix–xii.

65. Diana Long, "Physiological Identity of American Sex Researchers between the Two World Wars," in Geison, *Physiology in the American Context,* pp. 263–278.

66. Clarke, "Emergence," esp. pp. 154–165; Clarke, "Controversy." For discussion of legitimacy issues more generally, see Anselm L. Strauss, "Social Worlds and Legitimation Processes," in Norman Denzin, ed., *Studies in Symbolic Interaction,* vol. 4 (Greenwich, Conn.: JAI Press, 1982), pp. 171–190.

67. A. S. Parkes, "Prospect and Retrospect in the Physiology of Reproduction," *British Medical Journal,* 1962 (July 14): 71–75, esp. p. 72; and A. S. Parkes, "The Rise of Reproductive Physiology, 1926–1940, Dale Lecture for 1965," *Endocrinology* (Proceedings of the Society), 1966: xx–xxxiii, esp. p. xx.

68. Joan H. Fujimura, "Constructing Doable Problems in Cancer Research: Articulating Alignment," *Social Studies of Science,* 1987, *17:* 257–293; and idem, "The Molecular Biological Bandwagon in Cancer Research: Where Social Worlds Meet," *Social Problems,* 1988, *35:* 261–283.

69. Garland E. Allen, "The Introduction of Drosophila into the Study of Heredity and Evolution: 1900–1910," in Nathan Reingold, ed., *Science in America since 1820* (New York: Science History Pubs., 1976), pp. 266–277; Deborah Fitzgerald, "Tradition and Innovation in Agriculture: A Comparison of Public and Private Development of Hybrid Corn," in Lawrence Busch and William B. Lacey, eds., *The Agricultural Scientific Enterprise: A System in Transition* (Boulder and London: Westview, 1986), pp. 175–185; Diane Paul and Barbara Kimmelman, "Mendel in America: Theory and Practice, 1900–1919," in Rainger, Benson, and Maienschein, *The American Development of Biology,* pp. 281–310. For a comparison, see also Clarke, "Research Materials." On infrastructural requisites for research, see Adele E. Clarke and Joan H. Fujimura, eds., *The Right Tools for the Job in Twentieth Century Life Sciences* (Princeton: Princeton University Press, 1992).

70. Ross Granville Harrison, "Embryology and Its Relations," *Science,* 1937, *85:* 369–374, esp. p. 370.

71. See, for example, Rosenberg, "Factors"; and Mayr and Provine, *The Evolutionary Synthesis.*

72. See, for example, Hamilton Cravens, *The Triumph of Evolution: American Scientists and the Heredity-Environment Controversy, 1900–1941* (Philadelphia: University of Pennsylvania Press, 1978).

73. Jay L. Lush, "Genetics and Animal Breeding," in L. C. Dunn, ed., *Genetics in the Twentieth Century* (New York: Macmillan, 1951); Allen, "T. H. Morgan and the Split"; Barbara A. Kimmelman, "The American Breeders' Association: Genetics and Eugenics in an Agricultural Context, 1903–1913," *Social Studies of Sciences,* 1983, *13:* 163–204; Paul and Kimmelman, "Mendel in America"; Fitzgerald, "Tradition"; Daniel J. Kevles, *In the Name of Eugenics: Genetics and the*

Uses of Human Heredity (New York: Knopf, 1985); and T. H. Morgan et al., *The Mechanism of Mendelian Heredity* (New York: Holt, 1915).

74. P. E. Smith, "Experimental Ablation of the Hypophysis in the Frog Embryo," *Science,* 1916, *44:* 280–282; idem, "The Disabilities Caused by Hypophysectomy and Their Repair," *Journal of the American Medical Association,* 1927, *88:* 158–161; and idem, "Effect on the Reproductive System of Ablation and Implantation of the Anterior Hypophysis," in Allen, *Sex and Internal Secretions,* pp. 734–764; see also "Smith, Philip Edward," *Dictionary of Scientific Biography,* vol. 12 (New York: Scribner's, 1975), pp. 472–477.

75. The same was true in Britain. See, for example, Guy F. Marian, "Early Work on the Chemistry of Pregnanediol and the Oestrogenic Hormones: The Sir Henry Dale Lecture for 1966," *Endocrinology,* 1967, *35*(4): v–xvi.

76. Robert E. Kohler, "Medical Reform and Biomedical Science: Biochemistry: A Case Study," in Morris J. Vogel and Charles E. Rosenberg, eds., *The Therapeutic Revolution* (Philadelphia: University of Pennsylvania Press, 1979); idem, *From Medical Chemistry.*

77. Corner, *Seven Ages*; and Willard M. Allen, "Progesterone: How Did the Name Originate?" *Southern Medical Journal,* 1970, *63:* 1151–1155. H. L. Fevold, F. L. Hisaw, and S. L. Leonard, "The Gonad-Stimulating and the Leutinizing Hormones of the Anterior Lobe of the Hypophysis," *American Journal of Physiology,* 1931, *97:* 291–301; and Fulton and Wilson, *Selected Readings,* p. 403. Choh Hao Li, Miriam Simpson, and Herbert M. Evans, "Purification of the Pituitary Interstitial Cell Stimulating Hormone [LH]," *Science,* 1940, *92:* 355–356. Koch and his associates developed separation and distillation methods for hormones, especially in urine, and began synthesis of androsterone. Report for 1934–1935 on the Grant for Biological Research made by the Rockefeller Foundation, 8 November 1935, pp. 3–5. RF RG1.1, 5216d, Box 8, Folder 107, Rockefeller Archive Center. See also Fred C. Koch, "Biochemistry and Assay of the Testis Hormones," in Allen, *Sex and Internal Secretions,* pp. 372–391; idem, "The Male Sex Hormones," *Physiological Reviews,* 1937, *17*(2): 153; Allen and Doisy, "An Ovarian Hormone"; and see George W. Corner, Charles H. Danforth, and L. S. Stone, "Edgar Allen, 1892–1943," *Anatomical Record,* 1943, *86:* 595–597.

78. See, for example, Linda Gordon, *Woman's Body, Woman's Right: A Social History of Birth Control in America*, 2nd ed. (New York: Penguin, 1989); Kevles, *In the Name*; and James Reed, *The Birth Control Movement and American Society: From Private Vice to Public Virtue,* 2nd ed. (Princeton: Princeton University Press, 1983).

79. Borell, "Organotherapy and the Emergence," p. 1.

80. Division of Medical Sciences, National Research Council, *Contraception: Science, Technology and Application: Proceedings of a Symposium* (Washington, D.C.: National Academy of Sciences, 1979), p. v.

81. Funds were initially channeled through the Bureau of Social Hygiene and later through the Rockefeller Foundation itself. Aberle and Corner, *Twenty-Five Years of Sex Research,* p. 112–113.

82. Reed, *The Birth Control Movement,* p. 283.

83. For a thorough account of how the mission was redirected and how this fit well with the emerging funding goals of the Rockefeller Foundation, see Clarke, "Emergence," pp. 286–324. Aberle and Corner note that just after 1940, the com-

mittee returned to its original human sexuality research commitments when it provided major funding for the work of Alfred Kinsey; see Aberle and Corner, *Twenty-Five Years of Sex Research,* esp. pp. 92–100.

84. Aberle and Corner, *Twenty-Five Years of Sex Research,* esp. pp. 70, 92–100, and 199–200; see also Reed, *The Birth Control Movement,* pp. 311–366; and Philip J. Pauly, *Controlling Life: Jacques Loeb and the Engineering Ideal in Biology* (New York: Oxford University Press, 1987), pp. 182–200.

85. Robert E. Kohler, "The Management of Science: The Experience of Warren Weaver and the Rockefeller Foundation Programme in Molecular Biology," *Minerva,* 1976, *14:* 279–306, esp. p. 291.

86. Clarke, "The Emergence." For more general discussion of the development of scientific approaches to increased production in agriculture, see Charles E. Rosenberg, "Rationalization and Reality in Shaping American Agricultural Research, 1875–1914," in Nathan Reingold, ed., *The Sciences in the American Context: New Perspectives* (Washington, D.C.: Smithsonian Institution, 1979), pp. 143–163; Charles E. Rosenberg, *No Other Gods: On Science and American Social Thought* (Baltimore: Johns Hopkins University Press, 1976); Margaret Rossiter, "The Organization of the Agricultural Sciences," in Oleson and Voss, *The Organization of Knowledge,* pp. 211–248; and Margaret Rossiter, "The Organization of Agricultural Improvement in the United States, 1785–1865," in Alexandra Oleson and Sanborn C. Brown, eds., *The Pursuit of Knowledge in the Early American Republic* (Baltimore: Johns Hopkins University Press, 1976), pp. 279–298.

87. Pauly, *Controlling Life.*

88. Harrison, "Embryology and Its Relations," p. 370.

89. Willier and Oppenheimer, *Foundations.* Of the other two articles, one was by Charles M. Child on susceptibility gradients and the other by Hans Spemann and Hilda Mangold on the organizer concept; neither line of work was highly productive in the long run, although Spemann was awarded the Nobel prize in 1935. See Viktor Hamburger, *The Heritage of Experimental Embryology: Hans Spemann and the Organizer* (New York: Oxford University Press, 1988); and Horder, Witkowski, and Wylie, *A History of Embryology.*

90. Horder, Witkowski, and Wylie, *A History of Embryology,* pp. 111–112.

91. Allen, "T. H. Morgan and the Split," p. 123.

92. Ibid. See also Lush, "Genetics"; Barbara A. Kimmelman, "The Progressive Era Discipline: Genetics at American Agricultural Colleges and Experiment Stations, 1890–1920" (Ph.D. dissertation, University of Pennsylvania, 1987); idem, "The American Breeders' Association"; Fitzgerald, "Tradition"; and Harry A. Herman, *Improving Cattle by the Millions: NAAB and the Development and Worldwide Application of Artificial Insemination* (Columbia: University of Missouri Press, 1981).

93. See, for example, Gerald L. Geison, "International Relations and Domestic Elites in American Physiology, 1900–1940," in his *Physiology in the American Context,* pp. 115–154.

94. Clarke, "Controversy."

95. Jane Maienschein, ed., *Defining Biology: Lectures from the 1890's* (Cambridge, Mass.: Harvard University Press, 1986), p. 32; and idem, "History of Biology."

96. Clarke, "Controversy."

97. Evelyn Fox Keller, "Reproduction and the Central Project of Evolutionary Theory," *Biology and Philosophy,* 1987, *2:* 73–86; Elisabeth Lloyd, *All about Eve: Evolutionary Explanations of Women's Sexuality* (Princeton: Princeton University Press, 1990).

98. Allen, "T. H. Morgan and the Split"; and Jane Maienschein, "Preformation of New Formation—or Neither or Both," in Horder, Witkowski, and Wylie, *A History of Embryology,* pp. 73–108.

99. See, for example, the general thrust of Aberle and Corner's argument and Frank Lillie's "Classification of Subjects in the Biology of Sex," in Aberle and Corner, *Twenty-Five Years of Sex Research,* pp. 102–104.

100. Horder, Wikowski, and Wylie, *A History of Embryology,* p. 111.

101. Charles E. Rosenberg, "Toward an Ecology of Knowledge: On Discipline, Contexts and History," in Oleson and Voss, *The Organization of Knowledge,* pp. 450–455.

Hamilton Cravens

6

Behaviorism Revisited: Developmental Science, the Maturation Theory, and the Biological Basis of the Human Mind, 1920s–1950s

The years from the 1920s to the 1950s marked the halcyon days of interdisciplinary scholarship in American culture. In many fields of knowledge workers attempted to build intellectual and institutional networks with specialists in other fields. The common and guiding hope was that through the cooperation of fields that previously had been perceived as distinct and different, fruitful new ideas and consequences could be wrought, often from materials, notions, and circumstances that had in the preceding age seemed alien to one another. The new interdisciplinary or expansionist mentality of this thirty-year age reflected a particular way of understanding the character of natural and social reality, different indeed from the one that had preceded it—from the 1870s to the 1920s. From the point of view of that older taxonomy of natural and social reality, all the parts of the whole were segregated from one another—all were distinct from one another, or were said to be in opposition to one another. There was a whole that was the sum of these different parts, a system of discordances that made up a larger unity, but no more: the whole was a structure no greater than or different from the sum of its parts. In the perspective of the era of the 1920s to the 1950s, however, the whole was said to be different from or greater than the sum of its parts, and the parts of the whole were assumed to be distinct yet interrelated in that dynamic, functional unity. Indeed, it almost seemed as if oil and water would mix.[1]

Examples of the crystallization of the new mentality in the life sciences abounded in this new epoch. Consider, for instance, the shift in commonly accepted notions of inheritance. In the older view, heredity was particulate, constituted of specific hereditary units that stood for particular traits, as in Darwin's theory of pangenesis or Mendel's nonblending traits in the garden pea. And, as most biologists would have had it, heredity was entirely antagonistic to environment. In the new perspective, heredity and environment were understood as mutually interactive if also entirely different kinds of factors in the development of individuals and of species. The new revised Mendelian genetics of the interwar years assumed that individuals and species were dynamic, interactive entities, composites or statistical aggregates of many interactive factors that were more than or different from the sum of their parts. Similarly the older notion of species as a type in which all individuals fluctuated around a fixed mean, all but universal among biologists and life scientists, yielded to the new notions of speciation pioneered by population geneticists, in which species were understood as interbreeding populations whose members possessed a common core of dynamic, fluctuating genetic material. These new ideas soon set the stage for a widespread revolution in the agricultural industry, with the invention of new species, such as hybrid corn and rice, created for larger yields, or the "Davis tomato," made hard and square so as to fit into picking machines and shipping boxes, or genetically manipulated new breeds of livestock, such as the Butterball turkey and the so-called purebred Beefmaster cattle, to be larger and "better" flesh to eat, and of various chemicals to modify or replace natural processes, including such herbicides as 2,4-D, insecticides such as DDT, and growth stimulants, such as DES. Oil and water did indeed mix in such instances: the natural and the artificial, in a dynamic new unity.[2]

More than new interdisciplinary concepts and techniques emerged in these years. Whole new interdisciplinary fields took shape as well. Such fields were not without their prewar precedents, but in their new guise they took on different configurations. In the life or biological sciences an important example was child development, better known today as child and human development, or, more simply, developmental science. Child development grew out of the prewar child welfare movement, itself mainly the product of organized women's reform movements and, to a lesser extent, of the tiny band of child and animal psychologists in prewar America. But child development was essentially a postwar phenomenon, invented in no small measure by Rockefeller philanthropy. At the Laura Spelman Rockefeller Memorial in the 1920s, program officer Lawrence K. Frank brokered the expenditure of several million dollars to create an entire professional scientific subculture of child development, including a half dozen research centers, several score teaching and demonstration programs in child study and parent education, a scientific society, several technical journals, a popular magazine, *Parents,* and even a national postdoctoral fellowship program. During the hard times of the 1930s Frank worked for Rockefeller philanthropy again to help subvene his cherished interdisciplinary discipline, child development, this time at the General Education Board in charge of a gigantic, multi-institutional research program on the problems of youth and adolescence. Again several millions were spent on the field.[3]

Frank and the trustees for whom he worked always insisted upon an interdisciplinary orientation in the proposals that they funded in child development. Of course there was many a slip betwixt the cup and the lip. Not all scientists so funded necessarily did the kind of research that Frank might have wanted. In particular his brand of left-liberal Gotham politics tended not to be translated into the intellectual products of the mainstream workers in this new interdisciplinary discipline. Developmentalists happily cooperated with scientists in other fields so long as there was due recognition of priority and eminence. By and large most developmentalists were happy to take Frank's money—and ignore his politics, especially his dogma of the autonomous individual.

Developmental scientists in these years worked out a general theory of child and human development that was interdisciplinary in the sense that they used research from a variety of the biological as well as human sciences to buttress that theory. It had two parallel theses. The first of these was the notion of the innate intelligence quotient. Its champions insisted that mental growth and development were predetermined by inheritance, and therefore the IQ was fixed at birth. The other was the maturation theory. According to this argument growth itself was predetermined by the inheritance of the individual and by the phylogenetic endowment of the species or group to which the individual belonged. Thus maturation, not learning, explained development. As Florence L. Goodenough and John E. Anderson, professors at the University of Minnesota's Institute of Child Welfare, put it in their seminal text in the field, since "adult behavior is the outcome of earlier experiences operating upon native tendencies it cannot be understood without a knowledge of the stages through which it has passed."

Thus grand theory in developmental science was profoundly hereditarian. Goodenough and Anderson declared that the influence of the Johns Hopkins animal psychologist John B. Watson and his doctrines of behaviorism was immense on grand theory in developmental science and, in particular, on the field's maturation theory. As Darwin had made necessary a comparative psychology and Binet had invented a technique for measuring different levels of minds, they insisted, so Watson's investigations of infant behavior and of the modifiability or conditioning of emotional actions in early life were crucial. Thus, they argued, Watson had made possible the conceptualization of the origins of behavior as an integral part of the psychological network or larger entity.[4]

At first blush, the thesis that Watson's major scientific influence was to inspire hereditarian thinking appears bizarre. Watson's historical reputation as a radical environmentalist is well known. He did much to invent, or more precisely, to re-invent, himself as a diehard environmentalist and fearless critic of hereditarian thinking. And those who have written about him subsequently, whether psychologists or historians, have usually taken him at his word; more is the pity. His career as a scientist in the academic community was different from the path he trod in the 1920s.

In the 1920s, after he left academe, Watson took extreme environmental positions on child development in various public forums. This has led many to assume that his ultimate contributions to child and human psychology reflected those public stances. Yet neither Watson's ideas nor his legacy *as a scientist in the 1910s* were so one-sided as his later public statements might suggest. As Goodenough and Anderson testified, in the 1910s, before he left the Hopkins, Watson *and* his research had a more ambiguous, perhaps even different, influence than has been previously understood.[5] Watson's work and methods inspired hereditarian interpretations of human development. His successors in child development took from his work as a scientist that which they could adapt for their hereditarian schemes. Apparently the environmental determinism that some historians have seen in Watson's influence emanates from their own "presentist" perspectives, that is, a reading from B. F. Skinner in the 1950s (and other 1950s events and testimony) back the Watson of the 1920s, who was by then a reckless popularizer and environmental determinist. Such a view conveniently ignores too much. He had changed his ideas since leaving the academy.[6]

In 1914 Watson published *Behavior*, a state-of-the-art book in which he reviewed recent work on animal behavior. He believed that "the behavior man" should recognize no dividing line between man and brute in his objective study of behavior. As a thoroughgoing behavioral monist, he found much to criticize in the literature. Mind was everywhere and nowhere in the organism, he declared, insisting that "there are no centrally initiated processes." In this way he pushed his arguments as far as they could go. In effect, he got around the problem of the mind and of dualism by collapsing the distinction between body and mind. Body and mind constituted a dynamic, fluctuating unity constituted of many distinct yet interrelated elements. All behavioral responses were entirely physico-chemical in character, regardless of their simplicity or complexity. And all behavioral phenomena could be placed into one of three categories: sense-organ functions, instinctive functions, and learning. He rhapsodically proclaimed that in time behaviorists would trace the complete series of physico-chemical changes, or quantitative energy transformations, from the moment of the incidence of the stimulus to the end of the movement in the muscle. Thus Watson insisted that the stimulus awakened an innate or predetermined response in the organism.

For Watson reality was constituted of a whole larger than the sum of its parts. The organism was a complex, multi-organ, interdependent entity of distinct yet interrelated parts, including muscles, tissues, nerve cells, reflex arcs, and sense organs, that permitted the external stimuli and the internal inborn parts of the organism to interact on the three levels of the senses, the instincts, and learning or practice. Watson was neither a hereditarian nor an environmentalist. To him, both worked together, but in different ways at the three different levels of behavior. They were not opposite or antagonistic forces. They were mutually interdependent. They were part of a larger whole. All parts of that whole cooperated in a larger unity.

Watson also established the notion of stimulus and response, or the S-R mechanism. He argued that sense organs and instincts had an innate structure within the nervous system. Much among animals had a genetic or hereditary basis, and "the animal is born with certain systems of [reflex] arcs ready to function in serial order the moment the appropriate stimulus appears." Naturally since instinctive behavior patterns resulted when particular stimuli triggered them, Watson insisted, the whole organism was involved in the response. The nervous system, he continued, constituted a larger network or circuit. When a stimulus affected a receptor, there was an orderly progression of events, as much so as when later a habit was formed. The stimulus was "carried along performed and definite arcs to the effectors in the order in which the arcs offer the least resistance to the passage of the current." Thus in Watson's hands the stimulus-response mechanism became intensely rigid; all the external stimulus did was to "awaken" an innate "mute" biological structure.

Watson recognized learning as distinct from the mere maturation of innate behavior patterns among animals. He defined learning as the more complex behavior of the higher animals—songs in birds, hunting among mammalian predators, and so forth. Indeed, he even drew learning curves for different kinds of behavior patterns among various species to define the average time for an individual of a particular species to perform a particular act. With regard to the matter of the autonomy of the individual apart from the group (or species) to which he or she "belonged," Watson assumed, as did most scientists, that individuals varied only within the permissible range of their species. Thus he was interested in discovering the time the average cat or dog or horse would take to learn a particular action, which was not the same as the individual at all. Thus he thought in terms of group averages, not the variations of actual individuals. Ultimately the learning curves he derived reflected what he termed the interaction of learning and maturation, or environment and heredity. Yet no matter how much learning did occur, there was always a biological basis to animal minds in at least two senses: there was a genetic structure for all action; and all individuals could vary in their responses only within the limits of their species.[7]

In *Psychology from the Standpoint of a Behaviorist* (1919) Watson extended his interpretations of mind and behavior from animals to human beings. In the interval, he worked intensively with newborn infants in the John Hopkins Medical School Hospital (the first psychologist to do so) and he had come to understand that in infants, there were only three generalized reactions at birth: fear, love, and rage.[8] Watson distinguished between animals, which seemed to have many instincts, and humans, who apparently had few, mainly those behavior patterns evident in the neonate infant, such as nursing, yawning, or crying, or those in the infant to several weeks of age, including grasping and the Babinski reflex.

His overarching thesis that the organism considered as a whole was greater than the sum of its parts was evident in this book no less than it was with the book on animals. *Psychology from the Standpoint of a Behaviorist* was a comment on the field of human behavior as Watson understood it. In chapter after chapter he reviewed what was known—and what was not—about the human organism as behavioral machine. All of the parts of the organism's parts functioned together. Even so simple an activity as putting on clothes and lacing up shoes involved a highly complicated series of adjustments of a motor or glandular kind. And behavioral psychology was not merely physiology. Physiologists dealt with various parts of the organism that were physiological in character and function, whereas behaviorists, continued Watson, dealt "with adjustments of the organism as a whole." Thus in discussing innate or unlearned behavior patterns he emphasized how integrated they were with the entire organism. Emotions were an example of innate behavior patterns. They involved profound changes of the body as a whole—especially the "visceral and glandular systems"—but the body was a *system* of systems, dynamic, fluctuating, interactive, all different yet all bound together in infinite combinations and interactions.

Watson's interpretation of human thought was entirely consistent with his taxonomy of natural and social reality, his sense of the relationships between the whole and its parts. In which the whole was greater than or different from its components. The explicit and implicit language habits in adults were formed along with the explicit bodily habits and were bound up with them and became "a part of every total unitary system that the human organism forms." Language had an anatomical basis, he insisted. He listed the parts of the body that cooperated in the production of every spoken word, including the diaphragm, lungs, thorax muscles, larynx's extrinsic and intrinsic muscles, pharynx's muscles, nose, palate, cheek, tongue, and finally, the lips.

Watson pointedly refrained from mentioning the brain as an organ or body part as the physical or biological basis for human thought. That would have forced him to confront issues he had no wish to address. He even criticized the notion that language was a social, as distinct from a biological or natural, habit. If languages were inventions of people, and thus social products, hearing and speaking depended entirely upon the body's physical and biological organs or parts. Language habits were no more social than any other bodily habits in the way in which they were acquired by the individual as habits, he insisted. And thinking itself was "largely a verbal process," he declared. Thus Watson combined the natural and the cultural—or oil and water.

Then Watson discussed how the human organism worked as a total entity or network of various systems that cooperated together. "No matter what the human animal is doing," Watson insisted, "he works as a whole." The human personality was an entity greater than the sum of its parts, and it is "the result of what we start with and what we have lived through . . . the 'reaction mass' as a whole." he concluded. And of course the S-R formula was the mechanism through which the organism functioned. Most of that mass consisted of definite habit systems, instincts that in the normal human being had yielded to social control and emotions in the course of that person's life.[9]

In 1920 Watson had to resign his professorship at Johns Hopkins and to leave academic life because he had become involved in a sensational marital scandal.[10] Although he tried to keep his hand in science from his new position as an executive at the J. Walter Thompson Company, the famous New York advertising firm, his days as a working scientist were over. He did participate in New York's intellectual life, teaching at The New School for Social Research and engaging in debates over child-rearing and behaviorism until the later 1920s.[11] In these roles he took a far more radical environmentalist position than previously, as when he made his oft-quoted statement in *Behaviorism* (1925), that under certain broad conditions he could train any person to be any kind of adult: "give me a dozen healthy infants, well-formed, and my own specified world to bring them up in and I'll guarantee to take anyone at random and train him to become any type of specialist I might select—doctor, lawyer, artist, merchant-chief and, yes, even beggar-man and thief, regardless of his talents, penchants, tendencies, abilities, vocations, and race of his ancestors." Yet by then Watson was doing science as a glitzy Madison Avenue operator and he knew it. In the next sentence he admitted that he was going beyond his facts, but justified his action because those on the opposite side had done the same thing for centuries.[12] Eventually he withdrew totally into his new life as a highly successful advertising executive.[13]

Watson bequeathed a complex legacy to his successors in the emerging field of child development. He acquired the reputation for being a radical environmentalist after leaving academe. Such was the consequence of the extreme positions he took as a popularizer. Yet when he was still a professor his positions on these matters were far more sober, even restrained. His work then was neither hereditarian nor environmentalist in the sense of thinking of them as opposites, as was the general tendency among scientists before the nature-nurture controversy of the 1920s. He consistently invoked the *interaction* of heredity and environment, albeit with differing weights for the many behavior reactions he discussed. Thus he anticipated the nature-nurture controversy's resolution a decade or so before it took place.

More than many psychologists of his generation, he stressed the importance of the stimulus-response formula without the intervention of the mind or the brain as a central information-processing and manufacturing organ. Watson began the tradition of the S-R mechanism, rigid, unyielding, and automatic, in animal psychology and in child and human development. In his work with animal instincts, the importance of his contributions to ideas of growth and development can scarcely be overemphasized. He insisted that animal instincts matured or ripened with time, once they had been "awakened" by "appropriate" stimuli in the environment. This enabled him to avoid discussion of the brain or the mind, and to argue consistently that the organism as a whole, with its innate patterns and its learned responses, was involved in all behavior reactions. Indeed developmental scientists in general, and Florence Goodenough and John Anderson in particular, borrowed this formula from Watson. That most developmentalists gave Watson's model a somewhat different emphasis than he might have preferred was only to be expected.

Watson's work with infants was crucial to developmental science. He was the first American psychologist to do any kind of experimental work with infants and to publish the results of such experiments. In that sense his work was seminal. He had established the study of postnatal behavior for animals and humans. In his work with infants he had demonstrated the importance of conditioning for the shaping and reshaping of innate behavior systems, as in the experiments he and Rosalie Rayner (ultimately his second wife) performed in 1920 with an infant, to examine how fears could be induced and then transferred in humans.[14]

So committed was Watson to the American notion of growth and development from conception to adulthood that he never even considered the problem of deconditioning a child of fears once the child had been conditioned. This is a telling point for it never occurred to any North American scientist to work on the problem of deconditioning—devolution, or de-programming, as it were—before the middle 1920s, and even then there was only one instance before the 1950s.[15] Thus the little acorns to great oaks mentality, linear or evolutionary thinking in other words, was that powerful and self-evident a given to most Americans including Watson and developmental scientists. A direct corollary to such linear thinking was Watson's notion of learning, which he defined by group norms from which no individual within the group or species could deviate; he suggested as much when he drew up learning curves for average times for members of particular species to perform given tasks.

Thus Watson agreed with the new notions about speciation that biologists were beginning to formulate in the 1910s. That new model of a fluctuating, dynamic *population* has often been depicted as more "liberal" in its socio-political implications than the static one it replaced. While it is true that some individual scientists derived a scientific ideology from the new age's underlying taxonomy of reality that might be interpreted as "liberal"—the circle around the Columbia anthropologist Franz Boas springs to mind—it would appear in retrospect that most scientists (like most other Americans) deployed the taxonomy of reality's assumptions in a strikingly "conservative" manner, a reaction to be as expected from underlying struggles among competing interests in society and culture no less than the minority "liberal" response. It is important to note as well that in neither age—the era from the 1870s to the 1920s or that from the 1920s to the 1950s—were dominant notions of species, either as "types" or as "populations," recognition of individual development beyond the confines of the group (or race or species) to which the individual "belonged." Watson, like the vast majority of his scientific colleagues and fellow Americans, thought in deterministic ways about the development of individuals. All was predetermined; variation could take place only within the expectations of the type or population. Clearly Watson did his professional best to propagate such notions of individual development among his scientific colleagues in the 1910s, and to his various lay publics as a popularizer of science in the 1920s.

The key to Watson's influence on his successors was his prescience. In the 1910s he had anticipated the new taxonomy of natural and social reality of the era from the 1920s to the 1950s. In particular his blueprint of the whole as greater than or different from the sum of the parts became the model for the field, the notion of the organism as an entity greater than the sum of its parts, a network of inborn systems that responded to external stimuli without benefit of a central intellectual organ. This model of a dynamic system of systems, ever-changing, ever-unstable, ever-evolving, and constituted of infinite numbers of discrete yet interrelated parts, became pervasive in the biological and human sciences in the years between the two world wars. And scientists could deploy it whether they believed that causes of things came from a mixture of heredity and environment with the emphasis on the one or the other. And there was more. Watson also bequeathed to his successors the notion of the rigid S-R formula in which external stimuli aroused or awakened mute, inborn anatomical organs within the organism. Such a view of stimulus and response offered cold comfort indeed to champions of either environmentalism or indeterminism. It cannot be emphasized too strongly that Watson's public statements in the 1920s, after he left academe, had relatively little to do with his work as a scientist. That the neobehaviorists and antibehaviorists of the 1950s looked back to Watson as the founder of behaviorism should not blind us to Watson's actual positions as a scientist in the 1910s. His contemporaries and successors used his scientific work in ways uncongenial to his popular stances in the interwar years. About the only element of his pre-1920s work that he retained in his statements of the 1920s and beyond was the stimulus-response formula, in which the stimulus awakened genetically predetermined anatomical structures and "caused" behavior to happen. In the popular child-rearing book that he and Rosalie published in the later 1920s, for example, he argued that rigid, stimulus-response training would be the most efficacious to make children behave.[16]

By the later 1920s developmental science took shape institutionally and intellectually. These events awaited the completion of Lawrence K. Frank's grandiose schemes with the Laura Spelman Rockefeller Memorial's millions. He underwrote major expansion at the Iowa Child Welfare Research Station at the University of Iowa, and the Yale Institute of Psychology. And he oversaw the founding of new institutes of child welfare at Berkeley, Toronto, Minnesota, and Teachers College, Columbia University. It was not until then that the approximately 400 developmentalists in various disciplines ranging from anatomy to child growth and psychology to zoology began to acquire a sense of intellectual and disciplinary identity thanks to Frank's midwifery and related events such as conferences in the field sponsored by the National Research Council's Committee on Child Development.

Ultimately developmental scientists picked up where Watson had left off. They took his complex legacy and fashioned it into the maturation theory in the later 1920s. Two groups of specialists within developmental science were especially prominent in these efforts, animal psychologists and child psychologists. Both were interested in the same problem that Watson was: whether the "perfection" of instinctual and other kinds of early behavior resulted from learning and practice, and thus mainly from extrabiological factors, or whether it was the consequence of the maturation of internal systems within the larger whole such as the nervous system. Since most life scientists took the theory of evolution and the primacy of the animal analogy for granted in their studies of animals and humans, it was only to be expected that the findings of the specialists in animal behavior were regarded as fundamental and heuristic for those in human behavior.[17]

Starting in the 1910s, zoologists and psychologists began publishing studies of native behavior patterns in animals. In particular they were interested in the nature-nurture question. This work on animal instincts, of which Watson's work with animals was a notable and integral part, differed dramatically from the publications that American and British psychologists ever since William James had published on human instincts. Rather than merely proclaim the existence of instincts, as the writers on human psychology did, often with hilariously conflicting results, the workers on animal instincts tried to pose concrete questions about the development of patterns of behavior.[18]

Two psychologists at the University of Michigan, J. F. Shepard and F. S. Breed, established the genre of maturation versus learning or practice studies in a recognizably modern form in the early 1910s. They conducted experiments calculated to determine how much of the daily improvement in the pecking instinct of a variety of newborn chicks was due to practice and how much to maturation. They assumed that maturation and experience were different, even antagonistic, factors—oil versus water. Their assumptions were reflected in the experiments they designed, and in the experiments' results also. Shepard and Breed placed two lots of newborn chicks, twenty-three in five groups, in a dark room prior to the first tests, and prevented practice in pecking for differing amounts of time with each of the five groups—three, four, or five days after hatching. In the intervals they fed and watered the chicks by hand. Regardless of the duration of confinement, all chicks on the average began with a pecking efficiency of slightly less than 20 percent; with between one and two days' practice the chicks were up to "normal efficiency." Hence Shepard and Breed concluded that their curves for each of the five group averages showed that the first two days represented the contribution of continuous practice, and the rest of maturation, stimulated incessantly as it was by prior practice. In other words, practice was the stimulus of maturation. They were antagonistic forces that nevertheless worked in a larger system that was no greater than the sum of its parts. Thus Shepard and Breed accepted the older taxonomy of reality.[19]

If Shepard and Breed established the learning versus maturation formula as a general proposition, nevertheless their notion that continuous practice over a period of days stimulated maturation did not pass scrutiny among their professional peers. Watson, for one, had excoriated their work in *Behavior,* insisting that instincts in animals unfolded in serial fashion: a brief stimulus would awaken the predetermined anatomical structure, and maturation would take over. And Watson's notions of the genetic (i.e. deterministic) order of things, of the relations between the whole and the parts, differed drastically from that of Shepard and Breed. On either the specific or the general basis the work of Shepard and Breed did not sit well with their professional colleagues and rapidly became of historical interest only.

Watson had never outlined anything approaching a theory of maturation. He had comments on the matter aplenty, but no more. In the middle 1920s, scientists who were interested in the anatomical basis of behavior published highly influential work; these included the young animal psychologist Leonard Carmichael and the senior statesman of vertebrate anatomy George Ellett Coghill. They took the complex Watsonian heritage and worked out a theory of maturation in lower animals that became a model for workers in animal and child behavior, and thus in developmental science.

Carmichael did much to develop a general theory of maturation. Although he never claimed to be Watson's successor, by dint of his work and the vacuum of leadership in the field he became such. Educated at Tufts College and Harvard University with a brief seasoning at the University of Berlin, Carmichael published two articles while an instructor at Princeton University in which he outlined his general position on philosophical and methodological issues. In one, he attacked notions of psychological sensationism then current among psychologists for being grounded on such false assumptions as the separateness of sensations and the mind-body dualism. In the other, he staked out an interactionist position on the heredity-environment controversy. He insisted that the debate qua debate was futile and outmoded. Heredity and environment were clearly different kinds of forces. But the important point was that they worked together—they interacted, and the one was inconceivable without the other. They could not be segregated from each other even if they were different in essence. Both worked together in all processes of evolution, speciation, variation, and inheritance in the species, and all internal processes within the individual organism also. They were not antithetical, but distinct and yet interrelated parts of the larger whole.[20]

Over the next three years Carmichael conducted three experiments with frog and salamander embryos that enabled him to articulate a full-blown theory of maturation. In the first subseries Carmichael attempted to determine whether maturation was the consequence of innate and environmental factors working together. In these experiments he worked with control and experimental groups of the frog *(Rana sylvatica)* and the salamander *(Amblystoma punctatum)*. He extracted the eggs from the jelly for each of the four groups. He permitted the eggs for the control groups to develop in ordinary tap water. But he drugged the water for two experimental groups with the anesthetic chloretone in sufficient quantity so that the embryos would continue to grow but not be capable of any bodily movements. He permitted control groups to develop normally and carefully observed the articulation of the swimming reaction in both species. When five days had passed, he released the embryos in the two experimental groups from the anesthetic. Within no more than thirty minutes, all the embryos were responding to Carmichael's gentle external stimulation and were moving. Indeed, the average time for the embryos to shake off the effects of the anesthetic and to begin to move was twelve minutes. Within thirty minutes most of the embryos swam so well that only with difficulty could they be distinguished from the members of the control group who had been swimming for five days.

Carmichael insisted that this experiment failed to prove that maturation resulted solely from the action of innate factors. Reality was far more complex. The experiments did show that the embryos' reflex systems could function in biologically useful ways very quickly after they first began to move. But this did not mean that maturation was caused only by internal factors. Recent experimental workers on the neuromuscular system had demonstrated that its growth should be apprehended as a continuous living function. Thus the older conception of ontogeny as a process of the construction of a machine-as-organism that began to function as a living thing after it was constructed was fallacious. It would appear that living protoplasm functioned at all times. Thus development was an integrated and holistic process of dynamic, functional construction. Put another way, development was a process in which the organism was constantly changing from birth to death. Thus the intricate development of such interrelated structures as receptors, nerve trunks, central apparatus, and motor end-organs was probably caused by functional stimulation within the organism itself, a process that so intricately combined internal and external factors that there was no practical or theoretical way in which modern science could disentangle them. And indeed the external excitation and the internal response to the distinct yet interrelated elements of the neuromuscular system was itself an integral part of the larger growth process itself. And here it was obvious that there were definite stages in the processes of growth and development. Growth and development were orderly and sequential, not chaotic.

Thus in the experiments Carmichael performed with the embryos' maturing ability to swim, he insisted that the swimming movements were not perfected immediately. It took some time and therefore some development—and some practice as well—before the embryos could swim with full coordination. If coordination among individuals in each species varied, there were also limits to such variation, and therefore there were definite ranges and averages of variation for species as well as for individuals within each species. Carmichael declared this showed that biological factors did matter in maturation as well as external ones. Clearly there was no sudden emergence of the ability to swim. Hence the old idea of instinct as a performed idea, so common before the 1920s, was false. But this did not mean that inheritance and internal factors should be dismissed out of hand. Obviously both mattered. Neither could be extracted from growth, development, and maturation. Rhetorically he concluded, "Is development anything other than a process by which what is in the final analysis, an hereditary 'given,' is transformed by an environmental 'present'?"[21]

Carmichael used the same basic techniques to perform the other two experiments. In the second, he wanted to ascertain the time necessary to eliminate the effects of the anesthetic on the experimental embryos. New embryos were anesthetized, then awakened and permitted to swim freely for a day and reanesthetized for another day and reawakened again. He found no difference in the amount of time for the drug to wear off in both trials. He concluded that the time necessary for the appearance of the first movement was merely that for the drug to wear off and that the beasts were ready to move without much ado. That is, maturation was a process over time that was predetermined and extraordinarily powerful.

In the third experiment Carmichael wished to evaluate the importance of external stimulation. He wanted to establish whether external stimulation had to be continuous or whether one stimulus sufficed to "awaken" the organism's neurological system and to start its operations. Thus he was here attempting to resolve the difference of opinion between Watson and Shepard and Breed. He raised three groups of salamander embryos, one totally removed from light and sound, another in normal laboratory circumstances, and the third in a very noisy, well-lit room. There was no difference in the effect of the stimuli in the second and third groups. The same was true for those in the isolated group. They immediately responded to the stimulus of his flashlight when he entered the room and moved as if they were normal, responding right away to the first normal stimulus of light. Obviously continuous stimulation was unnecessary. One shaft of light awakened the genetically predetermined structures in the animals. Then they matured as their phylogenetic endowment dictated.

Carmichael insisted that the interdependency thesis explained all the facts. Maturation resulted from the interaction of external and internal factors. What we know about the operation of the nervous system, he argued, indicates that with no external stimulation, the response of the nervous system cannot take place, and there can be no neurological development or maturation and hence no behavior. The "internal stimulation and response of the nervous system must . . . be initiated by environmental stimulation," but once that stimulus had aroused the nervous system, development and maturation could continue for some time in "relative independence of external stimulation," he concluded on the basis of the second experiment and the extant literature in neurology. And by the same token, a neuromuscular apparatus that had never before functioned could determine the organism's external behavior (in this instance swimming) the very first time it was stimulated.

In Carmichael's notion of interdependence, the whole was greater than or different from the sum of the parts; thus he echoed Watson. And Carmichael also followed Watson in adopting a rigid stimulus-response formula. Both thought that heredity and environment worked together in development. They agreed far more than they disagreed in their interpretations of the development of animal behavior. Unlike Watson, Carmichael never did research on humans. Given the evolutionary analogy's power, and its oft-unspoken corollary, that what happened in animals was true for humans as well, Carmichael did not believe it necessary to work with humans. In a real sense he had done something even more insofar as he was concerned. He had worked directly on animals. And he had constructed a seemingly empirically sound, experimentally verified maturation theory that was to serve practitioners in child and human development for the next three decades.[22]

George Ellett Coghill of the Wistar Institute in Philadelphia had already provided much of the evidence and many of the arguments for the maturation theory. By the time Coghill came to the Wistar Institute in the mid-1920s, he had been a publishing scholar in neuroanatomy for almost a quarter century. The beast he selected or his lifelong labors was the salamander *Amblystoma*. Early in his labors he had interpreted the amphibian's growth and development from the older perspective, in which the emphasis was on the different parts of the larger organism, and a static view more generally. But in the 1910s, according to a former colleague, Coghill began to adopt the new taxonomy of reality, that is, growth and development of the nervous system from the standpoint of the integration of all into a larger unit. In 1928 scientists at University College, London, invited Coghill to give three public lectures on his work. Published as *Anatomy and the Problem of Behavior* (1929), Coghill's lectures became widely regarded by workers in the then-emerging field of developmental science as one of the most authoritative statements of the maturation theory.

Coghill had originally selected *Amblystoma* as his experimental animal because it was not overly specialized. Thus the beast could serve the purposes of general theory well. In *Anatomy and the Problem of Behavior* Coghill made his technical researches accessible to a general audience. Coghill's thesis was that the living body was an integrated unity from the beginning of life to the end and, therefore, various parts of the organism are under the control of the organism-as-a-whole. The living body was not a mere package of bits and pieces accumulated through accretion; such was an outmoded and erroneous perspective.

In his first lecture, Coghill made three general observations with regard to *Amblystoma*. First, the animal's behavior patterns developed in a regular, orderly sequence. That sequence was entirely consistent with the development of the nervous system and its parts. There was complete coordination between the inner nervous system and the external patterns of behavior. Thus total integration was the key to development. Second, Coghill continued, in this salamander physiological processes followed precisely the order of their embryological development in functions of movement in water, on land, and in feeding. And third, from the beginning behavior developed through the orderly expansion of a perfectly integrated total pattern. Within the larger whole, individuation of parts did take place. This was proven in the development of partial patterns with their own degrees of distinctiveness. Yet all distinct parts were integrated into the larger whole. Thus here Coghill concerned himself with the animal's behavior patterns and the explanations of those patterns on the basis of its functional nervous mechanisms.

In the second lecture, Coghill took up the problem of the development of the nervous system. The central issue was how did the conduction paths of the central nervous system acquire their definitive functions. The embryo was perfectly integrated before it had a central nervous system. The preneural system of integration thus overlapped the neural system in the regular course of its development. The organism-as-a-whole guided the development of the parts. In this instance the prior structure led the emerging additions and integrated them at the same time. All was fixed and established within the structure. In the first lecture, Coghill had addressed the issue of the nervous system as a conducting mechanism that determines the immediate behavior pattern, in the second, he had argued that the preneural functions overlapped the later-appearing and maturing neural functions in the development of the mechanism of behavior.

In the third lecture, Coghill took up the issue of how the preneural process of growth and differentiation in the nervous system, and especially in the individual neuron, participated in the functioning of the nervous system as a whole, as an integrated larger unity. In the development of behavior there were two processes taking place simultaneously. First, the total pattern expanded as a perfectly integrated unit. Second, as parts or partial systems grew they became differentiated with definite form, shape, and distinctiveness. From either perspective there was a dominant organic unity from the beginning. In conclusion, development of *Amblystoma* took place from the extension of the total pattern and not, as had been thought before, through the projection of separate and isolated parts that became integrated secondarily. Thus the fundamental function of the nervous system was to maintain the integrity of the individual.[23]

Thus Carmichael and Coghill had established the maturation theory's postulates with regard to lower animals. Many developmental scientists soon incorporated their researches. Coghill's published lectures, for example, quickly appeared on most reading lists and bibliographies in the field. Only rarely did the shy, remote Coghill attempt to spread his ideas through personal intervention. And he soon lost his institutional base. In the mid-1930s he was arbitrarily dismissed from the Wistar Institute and forced to retire to a small farm near Gainesville, Florida, where he eked out an exceedingly modest existence and died in 1941 of heart disease. Carmichael, on the other hand, rose to be one of the influential movers and shakers within the field of developmental science, serving as a major officer in several key societies and making widely regarded contributions to the field while teaching successively at Brown and Tufts before becoming secretary of the Smithsonian Institution after World War II. In the one instance the personal equation did not exist, in the other it did. Thus each was influential in his own way.

Both had taken Watson's legacy and turned it into a coherent theory of maturation. One could see the imprint of Watsonian thinking in that theory. The whole was greater than the sum of its parts: heredity and environment, and indeed all kinds of factors, continuously interacted. Behavior, not consciousness or other philosophical conceptions, was the proper study of humankind. And maturation was the ripening of the inner neurological structure and the brief, momentary awakening of each innate part thereof by its appropriate, momentary stimulus. For Carmichael no less than Coghill, the individual could vary only within the parameters of what its species' innate endowment and its environmental stimuli permitted (the latter selected by evolutionary forces). Over the next decade several investigators took up the problem of the pecking instinct in chicks again, and this time they followed the lead of Carmichael, Coghill, and what they had made out of Watson.[24]

Others were involved also as architects of the new theory of maturation. Above all, the actions and ideas of a small band of child psychologists were crucial. Some had been students of Clark University president G. Stanley Hall, who had cast child study as the study of the genetic development of the child. While that idea was not precisely the notion of maturation as articulated by Carmichael and Coghill and neuroanatomists more generally, clearly Hall's linear or longitudinal perspective and his great emphasis on the power of biological factors were highly congenial to the rapidly crystallizing maturation theory.[25]

In the 1920s child developmentalists launched several longitudinal or life-cycle research projects on children. The directors of these projects based their investigations on the assumption that the whole was greater than or different from the sum of its parts, and that child life should be studied in all of its infinite, diverse, interrelated, and dynamic characteristics. Hall's former doctoral students were prominent among these researchers.

Thus Lewis M. Terman, one of Hall's best-known students and the author of the Stanford-Binet intelligence test, initiated the first such self-conscious American longitudinal project on children in 1921. Terman strongly believed that heredity was mainly responsible for the levels of intelligence in individuals and in the groups in society and nature to which they belonged. Thus he embraced one of the leading theses of developmental science in the interwar years: the notion of the fixed IQ. He was especially interested in geniuses, believing them as individuals to be the movers and shakers of history and a nation's most precious resource. With support from the Commonwealth Fund he began collecting data with a team of assistants on 1,000 geniuses as identified by the Stanford Binet scale. Initially, in 1921, Terman seemed primarily interested in the role of mental inheritance in the making of geniuses. But he and his associates kept broadening the categories of the project so as to include more and more characteristics of the population they were attempting to describe, so that the blueprint so outlined was fully congruent with the taxonomy of natural and social reality of that age.

The Genetic Studies of Genius project, as Terman dubbed it, made a powerful contribution to the maturation theory in the 1920s and beyond. By 1930, three hefty tomes had appeared. In the first Terman and his associates published the results of their canvas of the mental and physical traits of the group of individuals they classified as geniuses. With his proclivity for assuming that innate high intelligence was a precious resource of civilization, it was not surprising that Terman wished the data to prove the general superiority of persons of high intelligence in all aspects of their personas and character, that is, their physical, emotional, moral, and social superiority, in addition to their intellectual quality. He had little difficulty demonstrating that the gifted California children in his sample were superior to the norms of the general American population, a fact which may have reflected the quality of life in early twentieth-century California as much as (or more than) the genetic or developmental character of genius as a natural fact. The second volume was a curious attempt to catalog the accomplishments of 300 geniuses in history and to estimate their IQs in retrospect from their known life histories. It was hardly astonishing that Terman found no mental midgets among them. In the third volume, published in 1930, the first five-year follow-up to the initial study, his techniques and assumptions had jelled absolutely. Perhaps not astonishingly Terman and his associates found that the geniuses were doing well in all respects. Terman attributed their success to their inheritance. Environment worked with heredity, to be sure, for it enabled the organism to exist and to work out its manifest destiny.

The studies were linear in that Terman and his associates followed a large group they defined over time, but the measurements and estimates they made of those in the sample were all cross-sectional portraits of the group as a whole, with individual persons studied only as examples of individual variation within the norms or means of the group as Terman and his associates thought appropriate. In other words, Terman and his colleagues never thought of tracking single individual geniuses (or persons, at any rate) over time. Rather they took "group portraits" at various intervals and thus had only group averages or means for all their measurements and, therefore, for all their generalizations. Strictly speaking, there was no such thing as an individual apart from the group of geniuses, or, in the abstract, of the individual apart from any kind of group in the study, and it was inconceivable to Terman and his associates that there should be. Terman's contribution to the notion of the fixed or innate intelligence quotient has been reasonably well known. What has not been so well understood have been the implications of his interpretation of the species question from biology for his conceptualization of the genius project. In effect, within the notions of growth and maturation as Terman enunciated them, an individual could not exist apart from the group to which "nature" (or "science?") had assigned him or her. Thus there was another regard in which the whole was greater than any or all of its parts. Here Terman, wittingly or not, had teased out the implications of Watson's prescient notion of the relationship between the individual and the species.[26]

Another of Hall's former students who did seminal life-cycle research in the 1920s and beyond was Arnold L. Gesell of Yale. Gesell did not study an experimental population, after the fashion of his rival Terman. Rather he took as his research population those children whose parents brought them to his Psycho-Clinic at the Yale Medical School for consultations, apparently without consideration for any flaws in his methods of sampling. Gesell sought to study the phenomenon of maturation as evidenced in early human growth, mental and physical maturation. So for him the individual and the group stood for each other just as they did in a very different manner for Terman—and the most other scientists of the day. In several early works, including *The Mental Growth of the Preschool Child* (1925) and *Infancy and Human Growth* (1928), Gesell outlined the processes of maturation and growth, which he patterned after the ideas of Carmichael, Coghill, and other animal developmentalists. All the ultimate unity of the organism was reflected in the differentiation and maturation of the various parts of the nervous system and the body, he argued. This was the great principle of maturation: there was total integration. Gesell himself was not as interested in the heredity-environment issue as was Terman, in part because it was largely irrelevant to Gesell's work. Gesell had little investment in intelligence tests as Terman did. And Gesell's model of maturation assumed the interaction of nature and nurture in the individual and in the species, but also projected the integration of all the diverse characteristics of the individual as maturation took place from conception to adulthood. The whole was thus greater than the sum of its parts.[27]

Other child psychologists in the later 1920s and early 1930s conducted less ambitious research projects on the phenomenon of maturation. Thus Mandel and Irene Case Sherman made a preliminary investigation in 1925 at the Northwestern University Medical School of whether it was possible to make a quantitative study of the sensorimotor responses of newborns. They studied ninety-six infants, ranging in age between one hour and twelve days. The particular responses that the Shermans examined were the plantar and pupillary reflexes, responses to being pricked with a needle, early habit formation of defense movements of the arms, and the coordination of the eye muscles. It should be possible, declared the Shermans, to make a quantitative study of such phenomena to see how close they are to perfect at birth, their rate of improvement, and the age at which they were perfected. In this particular investigation, all "sensori-motor responses were found to be imperfect at birth, and showed an increase in adequacy with the advance in age, up to a certain point, at which the responses were perfected."[28]

At the University of California's Institute of Child Welfare, psychologists Harold E. and Mary Cover Jones conducted an experiment to test the role of maturation in altering emotional patterns. The Joneses had known Watson well in New York. Thus their experiment might be regarded as an extension of Watsonian thinking on such matters into the later 1920s. The Joneses exposed some fifty-one children in the institute's nursery school and laboratory to a harmless black snake, in small groups and by surprise. Children up to two and a half years old had no fear whatsoever, whereas children three and a half years old were commonly cautious, children four years and older were afraid, and fear was definitely quite pervasive among the ninety-one adults exposed to the snake as compared with the young children. The Joneses insisted that there were only three possible explanations for the facts so accrued, namely that they could have resulted from conditioning, from the ripening of the innate instinct or fear of snakes, or, third, from the general maturation of behavior, which led to greater sensitivity and more discriminatory responses. The Joneses rejected the first explanation because there had been no conditioning of the subjects. The second explanation, a special instinctive fear of snakes, they continued, "is related to the doctrine of innate ideas and the inheritance of acquired characters, and is not in keeping with present-day theory." The third possibility, they declared, had to be the answer. As the child matured, its nervous system made it possible to have more differentiated and specific responses and evaluations of situations. Fear obviously arose when individuals knew enough to recognize the potential danger in the situation but were not sufficiently advanced to the point of complete comprehension of the situation; that is, the harmlessness of the black snake, which information even the adults did not securely possess.[29]

Over the next several years a number of investigators conducted more experiments on maturation. All were similar methodologically in that control and experimental groups were used to discriminate between "maturation" and "practice." In two parallel studies of the role of maturation versus practice or learning, Arthur I. Gates and Grace A. Taylor of the Institute of Child Welfare at Teachers College, Columbia University concluded that no amount of practice could give individuals who had not matured any permanent or lasting advantage in either motor or mental functions.[30] Florence L.

Goodenough and C. R. Brian at the University of Minnesota's Institute of Child Welfare came to similar conclusions using the same method of control groups, which had no practice, and experimental groups, which did, in the particular skill they were testing.[31]

Probably the most influential of these maturational studies was that of Arnold Gesell and his associate Helen Thompson. They studied two identical twins. Each twin acted as both a control and a practice on alternating sets of skills for the other. Gesell and Thompson argued that beyond doubt maturation was far more important than practice or learning in the sense that the infants could not do certain tasks until their nervous systems had matured to the point at which they were capable of performing the tasks so indicated.[32] In a similar study published several years later, the young Yale psychologist Josephine R. Hilgard studied the same twins as Gesell and Thompson, now somewhat older, in various tests of motor skill coordination, one as a practice and the other as a control, for alternating tests. Her conclusions were virtually identical with theirs. Although each twin acquired different attainments, both returned to similar levels of performance when practice sessions ended. They had to wait for general maturation to finish before these attainments became permanent.[33] And that was the general position of the maturational studies done in these years: that performance or skill depended on the maturation of the central nervous system, on its integration of many distinct parts into a larger whole, just as Watson, Carmichael, and Coghill had argued with varying degrees of emphasis.[34]

All of these maturational studies focused on the development of motor skills and abilities. To most workers in the field, the assumption of the fixed intelligence quotient in individuals addressed the problem of the maturation of the intellect, which all agreed was an entirely different kind of phenomenon than motor coordination and skill. Another step in the articulation of the maturation theory was taken when someone provided a direct link between the two assumptions—of the regularity of innate development and of the fixity of intelligence. At the University of California's Institute of Child Welfare, Nancy Bayley, a postdoctoral research fellow, began in 1928 what was to become one of the institute's three celebrated longitudinal research projects, the so-called Berkeley Growth Study. Bayley selected a group of seventy-four newborn infants from the community, chiefly white and middle class. Initially Bayley intended the study to be limited to the first year of life, and to be a study of all aspects of infants and their behavior on the assumption that the organism was an entity larger than or different from the sum of its parts. But once the project got under way, funding from Rockefeller philanthropy ensured steady work, and Bayley remained with the project until her retirement as a postdoctoral fellow at the university more than three decades later.

In 1933 Bayley published a monograph in which she reported on the mental growth of the remaining children in the project. At this time and until the 1950s, Bayley took for granted the notion of regular, predictable maturation in all aspects of growth and development, including intellect, and, like most every other developmental scientist, she assumed that individuals did not vary beyond the "natural" range of their group. Here she was concerned with testing the hypothesis, which the British psychologist Charles Spearman had advanced in the early 1900s, that there was in the human endowment a general factor (or "g") that stood for general intelligence. Scientists on both sides of the Atlantic continued to take Spearman's theory as gospel well into this century. Indeed, many had assumed that Mendel's laws of heredity corroborated Spearman's idea of "g".

Bayley found that the Spearman theory of a single factor for intelligence was simply erroneous. In doing so she explicitly welded the two maxims of grand theory in developmental science: the notion of the fixed IQ and the maturation theory. Others such as Terman had assumed the coequal status of the two ideas, but it was Bayley who did the formal honors with her monograph. She wrote from the perspective of that age's distinctive notion of the taxonomy of natural and social reality. The many tests she deployed with her charges measured many different functions and showed that mental life and its developmental and constitutional bases were far different and much more complex.

Bayley found mental growth to be a complicated phenomenon. Growth was very rapid in the early months, with a deceleration after about ten or eleven months, but after fifteen months the rate increased and was almost constant. With growth came variability within the group, and as the children grew older they became more variable in their successes at all age levels. There was no consistency in the children's test scores over a long period, but the scores of "adjacent" tests correlated rather highly. Development in the first six to eight months was largely sensorimotor in character. The more truly adaptive behavior emerged (and was tested) later. For the first seven months, Bayley continued, there were negative statistical correlations between the children's scores and the educational levels of their parents, but with time the correlations became positive. At every point, Bayley derived, used, and compared *group* averages, not the scores of *individuals*.

Withal Bayley thought of the project and its results from the standpoint of the ultimate assumptions of maturation theory and the fixed IQ. Thus her work became, when considered with that of her like-minded colleagues, doubly reinforcing. That is, belief in the maturation theory seemed to corroborate the notion of innate intelligence, and vice versa. Yet in reality it was her tacit belief in the taxonomy of natural and social reality that the whole was greater than or different from the sum of the parts that bound the two themes of the fixed IQ and predetermined development together. Accordingly, there were no direct links of factual evidence binding the ideas—just the assumptions themselves.[35] And what was true of Bayley was true of most co-workers in the field until the 1950s.

At least as revealing a manifestation of the emergence of the maturation theory among professionals in the field was the publication of Carl Murchison's *Handbook of Child Psychology* in two editions, in 1931 and 1933. Murchison's *Handbook* was not interdisciplinary. It was a compendium of state-of-the-art essays on special topics within child psychology with no attempt to address related fields (anatomy, pediatrics, and nutrition) which Lawrence K. Frank had tried to orchestrate in his grandiose funding plans. Yet even in its most interdisciplinary format, as in the Conferences sponsored by the National Research Council's Committee on Child Development in 1925, 1927, 1929, and 1933, it was clear that child psychology—the behavior and mind of the child in all of its manifestations—was the inner core of the field of child development.[36]

In the first edition of his *Handbook*, Murchison solicited essays from experts in the field on special topics that warranted special attention. As he remarked, the field was already astonishingly filled with solid accomplishments. Unfortunately, "many experimental psychologists . . . look upon the field of child psychology as a proper field of research for women and for men whose experimental masculinity is not of the maximum," a patronizing attitude based on "a blissful ignorance of what is going on in the tremendously virile field of child behavior," Murchison declared. Thus manful pride in the field seemed to have animated Murchison. He was, after all, successor to G. Stanley Hall at Clark and thus was professionally committed to the advancement of Hall's favorite topic, child psychology.

Murchison addressed the volume to professional psychologists who wished to know more about this rapidly growing field. There were twenty-two authors and essays in the volume, on a wide variety of topics, including the conditioning of emotions, on drawings, on plays, games, and amusements, on psychoanalysis, on morals, and on primitive children, but also such explicitly natural-science problems as learning, developmental psychology in twins, physical and motor development, language development, and eating, sleeping, and elimination, and such special problems as the gifted child, feeblemindedness, and other special gifts and special deficiencies.

Worth noting in the first edition was that the chapters did not advance from one set of developmental problems to another. Murchison thought of the field as a series of particular problems. Those chapters that had intellectual affiliations with the natural sciences Murchison did place before those that could be identified with the applied social sciences, psychiatry, and clinical psychology. If Murchison had an underlying theme, that was it. Furthermore, authors did not adhere to one particular set of ideas about child and human development. There was as yet no settled dogma, no pure truth—just a surprising amount of good work, as Murchison thought of the matter. For instance, there were hereditarians such as Arnold Gesell, Florence Goodenough, and John E. Anderson, who insisted that nature prevailed over (as well as interacted with) nurture, but also those who argued for the importance of the cultural environment, whether immediate, such as Kurt Lewin, or historic, such as Margaret Mead, or that of the special preschool for disadvantaged youngsters, such as Susan Issacs. Further evidence of Murchison's catholicity came with his inclusion of such authors as Anna Freud, Jean Piaget, and Lewis M. Terman, who could hardly be said to agree with one another (and other authors as well) on much of anything in the field save that it should exist as a field of investigation.[37]

In 1933 Murchison brought out the revised second edition. As he acknowledged, this edition was very different. Much had changed in the past three years. The field itself had expanded greatly, especially "the renewed interest in the study of prenatal behavior and the behavior of the neonate," and other topics had disappeared from the volume because they were no longer under research. The most important difference between the first and second edition flowed from Murchison's clue of the new work in prenatal and neonate behavior: the second edition was now organized, and read, like a manual of the various topics or problems related to maturation, including maturation itself. In brief, it was intellectually organized as a text of the maturation theory.

Thus Carmichael had the lead substantive chapter, 128 pages in all, including references to twenty-four of Coghill's more important publications, so that Carmichael spun out the maturation theory, introduced fellow developmentalists to Coghill's work, and provided them with a total of 354 European and American references in neuroanatomy and related fields. Then came several more new chapters that further reinforced the maturation theory, including one by Gesell on maturation and the patterning of behavior, and others on the neonate and on locomotor and visual-manual functions. Florence L. Goodenough added a new chapter on mental measurement, the thesis of which was that the IQ was fixed at birth. She was not the only author to affirm the truth of this maxim, but she was easily the most resolute. There were also several more chapters from the first edition included with the new ones, but now arranged so that they made sense from a developmental perspective. Then came two more groups of chapters—one on factors that modified child behavior, the other on special groups of children—with about half of them new. One could see the child develop from conception to adulthood in all of his or her complexity. It was as if the very order and arrangement of the chapters of the book itself illustrated the development of the child as a phenomenon of nature just as the scientific essays contained therein "proved" the twin hypotheses of genetically predetermined maturation and the fixed IQ as the results of scientific investigations.

Furthermore, there was another and more subtle change: with the major exception of Kurt Lewin's chapter on environmental forces acting on the child, the *Manual's* other chapters read as if all behavior arose from the awakening of predetermined mute anatomical structures from a single stimulus in the developmental cycle. The Watsonian imprint was indeed there. The whole was indeed greater than or different from the sum of the parts, and internal and external systems interacted in all behavior of individuals and of species. Furthermore, the Watsonian stimulus-response formula was indeed an integral part of explanations of development and behavior as those issues arose. Many authors cited both John B. and Rosalie Rayner Watson in both editions. And it was clear that Watson had a legacy, an influence in the field, long after he had left it for the delights of Madison Avenue albeit not quite the one he might have imagined. The authors in the first edition barely recognized Coghill or Carmichael, qualitatively or quantitatively. In the second edition the situation was quite the reverse. Coghill, Carmichael, and Watson were widely noted. The maturation theory was pervasive in the book, even more so than the notion of the fixed IQ, for the good and simple reason that the constancy of the IQ was taken for granted by most psychologists. Now the concrete details of the about-face Murchison made from the first to the second volume may never be known. But clearly no editor could avoid pressure from specialists to take cognizance of new ideas that they thought compelling and true.[38]

With the publication of Murchison's second edition, the maturation theory had won important recognition and legitimacy among developmentalists. If there were space here to flesh out the further details of maturational studies, it would be largely a story of repetition until the late 1940s and early 1950s, when developmentalists suddenly began asking new questions from different perspectives. We can note, however briefly, that the year 1933 was a crucial one in the history of developmental science. The appearance of Murchison's second edition signaled the ascendancy of the maturation theory among developmentalists as defined here. And the founding of the Society for Research in Child Development made 1933 doubly important, at least symbolically. It was organized under the aegis of the Committee on Child Development of the National Research Council. It gave the developmentalists a scientific society all their own and, following the notions of that age, an explicitly interdisciplinary one at that. Now developmentalists regardless of disciplinary affiliation could participate in the organization, and thus in the field. The SRCD carried on a vigorous publication program.[39]

Probably the two men who did the most in the field to represent the maturation theory and to endorse it through theory and experiment were Arnold Gesell and Leonard Carmichael. After 1933 Gesell suffered few intellectual or theoretical challenges. He was an effective popularizer for the maturation theory.[40] As for Carmichael, when the time came after the Second World War for a new edition of Murchison's *Manual* to be issued, it was he who stepped in and became organizer and editor, with intellectual results that hardly bear repeating here.[41] For a time, then, the field had jelled intellectually. Indeed, to some in the 1940s it seemed as if it were almost a static field in which all the principles and theories were well understood. That was, of course, merely the calm before our contemporary storm of individualism and individualistic fragmentation, which began in the 1950s. But that is another historical problem.

Was the maturation theory simply an arcane idea of a few hundred scientists from 1920 to 1950? This is not the place to delve into that question, but it might be remembered that the popular American belief in upward social mobility in that age was that it was possible for groups to change position, but not for individuals to jump from one group with low standing to another with a higher reputation; or, as it was often put, a rising tide carried all boats. That homely piece of folk wisdom affirmed that there was no such thing as an individual out of place in nature or society, for everything fit with everything else—even oil and water. Any idea to the contrary would have been inconceivable, or, if conceivable to some, then incredible to all others.

Acknowledgments

The author thanks the coeditors and coauthors of this volume for their constructive criticism of an earlier draft of this essay and assumes all responsibility for any remaining errors therein. He also thanks the Office of the Vice Provost for Research, Iowa State University for material support in the preparation of this essay, as well as the American Society of Zoologists for partial reimbursement for air travel to meet with colleagues at Friday Harbor, Washington. He would like to thank his colleagues in his department's discussion group for helpful suggestions as well. He is also grateful to the staffs of the libraries of Georg-August University, Goettingen, Bundesrepublik Deutschland, for their cooperation in obtaining materials while he wrote the essay as George Bancroft Professor of American History at the university.

Notes

1. There were a few notable biologists active before the 1920s who anticipated the post-1920s taxonomy of reality, including Charles Darwin, especially his "tangled bank" metaphor in *Origin of Species* (1859) and Claude Bernard, in his classic *An Introduction to the Study of Experimental Medicine* (1865). But they were

exceptional, to be fully understood by scientists in the 1920s. For a more general discussion, see Garland E. Allen, *Life Science in the Twentieth Century* (New York: Cambridge University Press, 1978), passim. A sophisticated case study of this transformation of worldviews is Alan I. Marcus, "From Ehrlich to Waksman: Chemotherapy and the Seamed Web of the Past," in Elizabeth Garber, ed., *Beyond the History of Science. Essays in Honor of Robert E. Schofield* (Bethelehem, Pa.: Lehigh University Press, 1990), pp. 266–283.

2. See, for example, Hamilton Cravens, *The Triumph of Evolution. The Heredity-Environment Controversy, 1900–1941* (Baltimore: Johns Hopkins University Press, 1988 [1978]), pp. vii–xvi, 15–55, 157–190, and passim. Alan I. Marcus and Howard P. Segal, *Technology in America: A Brief History* (San Diego: Harcourt Brace Jovanovich, 1989), pp. 275–283, has a penetrating discussion of the "revolution" in agriculture in these years.

3. Robert R. Sears, *Your Ancients Revisited: A History of Child Development* (Chicago: University of Chicago Press, 1975); Hamilton Cravens, "Child-Saving in the Age of Professionalism, 1915–1930," in Joseph M. Hawes and N. Ray Hiner, eds., *American Childhood* (Westport, Conn.: Greenwood Press, 1985), pp. 415–488; idem, "The Wandering IQ: Mental Testing and American Culture," *Human Development,* 1985, *28:* 113–130; idem, "Recent Controversy on Human Development: A Historical View," *Human Development,* 1987, *30:* 325–335.

4. Florence L. Goodenough and John E. Anderson, *Experimental Child Study* (New York: Century Company, 1931), p. 17. Goodenough and Anderson were thoroughgoing hereditarians. And, indeed, the text Goodenough and Anderson published became a major instrument in the dissemination of Watson's infant researches to the next generation of developmentalists.

5. On behaviorism, see: John C. Burnham, "On the Origins of Behaviorism," *Journal of the History of the Behavioral Sciences,* 1968, *4:* 1143–151, which emphasizes Watson's role as a charismatic leader in his field when he published his supposedly epochal 1913 manifesto of behaviorism; Cravens, *The Triumph of Evolution,* pp. 201–210, treats Watson's work only to 1917 and as catalyst to the instinct controversy, but argues elsewhere that the mainstream ideas coming from the nature-nurture controversy were deterministic and naturalistic, as well as strongly hereditarian; Franz Samelson, "Struggle For Scientific Authority: The Reception of Watson's Behaviorism, 1913–1920," *Journal of the History of the Behavioral Sciences,* 1981, *17:* 399–425, persuasively insists that Watson's ideas and methods had little discernible impact among psychologists in the 1910s; Robert Boakes, *From Darwin to Behaviourism, Psychology and the Minds of Animals* (Cambridge: Cambridge University Press, 1984), pp. 143–175, 218–241, accurately discusses Watson and his work, and competently treats behaviorism and its influence to about 1930; John M. O'Donnell, *The Origins of Behaviorism, American Psychology, 1870–1920* (New York: New York University Press, 1985) is a fine social history of American psychology to the 1920s which does not (and need not) address the issues raised in this essay. A recent biography of Watson is Kerry W. Buckley, *Mechanical Man, John Broadus Watson and the Beginnings of Behaviorism* (New York: Guilford Press, 1989).

6. A recent example of this flawed presentist perspective is Robert J. Richards's

prize-winning *Darwin and the Emergence of Evolutionary Theories of Mind and Behavior* (Chicago: University of Chicago Press, 1987), pp. 504–511.

7. Quotations from John B. Watson, *An Introduction to Comparative Psychology* (New York: Holt, 1914), pp. 18, 45–46, 53, 148, 259–260, and passim. See also Watson, "An Attempted Formulation of the Scope of Behavior Psychology," *Psychological Review,* 1917, *24:* 329–352; idem, "A Schematic Outline of the Emotions," *Psychological Review, 1919, 26:* 165–196.

8. See, for example: John B. Watson to Robert M. Yerkes, 27 October 1915, 12 October 1916, 18 October 1916, Robert M. Yerkes to John B. Watson, 16 October 1916, John B. Watson Author Folder, Robert M. Yerkes Papers, Library, Yale University, New Haven, Connecticut (hereafter Yerkes Papers).

9. John B. Watson, *Psychology from the Standpoint of a Behaviorist* (Philadelphia: Lippincott, 1919), pp. 213, 214, 330, 331, 342, 355, 369, 440, and passim.

10. Boakes, *From Darwinism to Behaviourism. Psychology and the Minds of Animals,* p. 224, discusses the matter.

11. Peter M. Rutkoff and William B. Scott, *New School. A History of the New School For Social Research* (New York: Free Press, 1986), p. 36.

12. Quotations from John B. Watson, *Behaviorism* (New York: Norton, 1925), p. 104.

13. On Watson's own adjustment to the business world, see, for example, John B. Watson to Robert M. Yerkes, 7 January 1922, 2 May 1923, 15 August 1923, 19 January 1926, 21 January 1926, 17 February 1926, 10 May 1926, 13 May 1932, John B. Watson Author Folder, Yerkes Papers.

14. John B. Watson and Rosalie Rayner, "Conditioned Emotional Reactions," *Journal of Experimental Psychology,* 1920, *3:* 1–14. See also Ben Harris, "Whatever Happened to Little Albert?" *American Psychologist,* 1979, *34:* 151-160.

15. Boakes, *From Darwin to Behaviourism. Psychology and the Minds of Animals,* pp. 222–223. This is an excellent book with many penetrating insights. The exception to the rule was Mary Cover Jones, "A Laboratory Study of Fear: The Case of Peter," *Journal of Genetic Psychology,* 1924, *31:* 308–315.

16. John B. Watson and Rosalie Rayner Watson, *The Psychological Care of Infant and Child* (New York: Norton, 1928), passim. For Rosalie's dissent from her husband's rigidity, see Boakes, *From Darwin to Behaviourism, Psychology and the Minds of Animals,* pp. 227–228.

17. Cravens, "Child-Saving in the Age of Professionalism, 1915–1930," pp. 440–452.

18. Calvin P. Stone, "Recent Contributions to the Experimental Literature on Native or Congenital Behavior," *Psychological Bulletin,* 1927, *24:* 36–61. On the vogue of human instincts, see Cravens, *The Triumph of Evolution. The Heredity-Environment Controversy, 1900–1941,* pp. 71–78; Hamilton Cravens and John C. Burnham, "Psychology and Evolutionary Naturalism in American Thought, 1890–1940," *American Quarterly,* 1971, *23:* 635–657; and Luther Lee Bernard, *Instinct. A Problem in Social Psychology* (New York: Macmillan, 1924), which catalogs the many lists of human instincts in this particular genre.

19. J. F. Shepard and F. S. Breed, "Maturation and Use in the Development of Instinct," *Journal of Animal Behavior,* 1913, *3:* 274–285.

20. Leonard Carmichael, "An Evaluation of Current Sensationism," *Psychological Review,* 1925, *32:* 192–215; idem, "Heredity and Environment: Are They Antithetical?" *Journal of Abnormal and Social Psychology,* 1925, *20:* 245–260.

21. Leonard Carmichael, "The Development of Behavior in Vertebrates Experimentally Removed from Influence of External Stimulation," *Psychological Review,* 1926, *33:* 51–58.

22. Leonard Carmichael, "A Further Study of the Development of Behavior in Vertebrates Experimentally Removed from the Influence of External Stimuli," *Psychological Review,* 1927, *34:* 34–47; idem, "A Further Experimental Study of the Development of Behavior," *Psychological Review,* 1928, *35:* 253–260. For Carmichael's later work, see, for example: Carmichael, "An Experimental Study in the Prenatal Guinea-Pig of the Origin and Development of Reflexes and Patterns of Behavior in Relation to the Stimulation of Specific Receptor Areas During the Period of Active Fetal Life," *Genetic Psychology Monographs,* 1934, *16:* 337–497; idem, "A Re-Evaluation of the Concepts of Maturation and Learning as Applied to the Early Development of Behavior," *Psychological Review,* 1936, *43:* 450–470.

23. G. E. Coghill, *Anatomy and the Problem of Behavior* (Cambridge: At the University Press, 1929). See also Coghill, "The Genetic Interrelation of Instinctive Behavior and Reflexes," *Psychological Review*, 1930, *37:* 264–266. A most interesting book about Coghill is C. Judson Herrick, *George Ellett Coghill. Naturalist and Philosopher* (Chicago: University of Chicago Press, 1949), passim. I am indebted to Professor Sharon Kingsland for bringing this book to my attention.

24. Charles Bird, "The Relative Importance of Maturation and Habit in the Development of an Instinct," *Journal of Genetic Psychology,* 1925, *32:* 68–91; idem, "The Effect of Maturation upon the Pecking Instinct of Chicks," *Journal of Genetic Psychology,* 1926, *33:* 212–233; idem, "Maturation and Practice: Their Effects upon the Feeding Reactions of Chicks," *Journal of Comparative Psychology,* 1932, *16:* 343–366; Wendell Cruze, "Maturation and Learning in Chicks," *Journal of Comparative Psychology,* 1935, *19:* 371–409; idem, "Maturity and Learning Ability," *Psychological Monographs,* 1938, *50*(5): 49–65.

25. The standard biography of Hall, which discusses his role in the promotion of child study and his students, is Dorothy Ross, *G. Stanley Hall. The Psychologist as Prophet* (Chicago: University of Chicago Press, 1972).

26. Lewis M. Terman, et al., *Genetic Studies of Genius,* vol. 1: *Mental and Physical Traits of a Thousand Gifted Children* (Stanford: Stanford University Press, 1925), passim; Terman and Catherine M. Cox, vol. 2: *The Early Mental Traits of Three Hundred Geniuses* (Stanford: Stanford University Press, 1926), passim; and Terman, B. S. Burks, and D. W. Jensen, vol. 3: *The Promise of Youth: Follow-Up Studies of a Thousand Gifted Children* (Stanford: Stanford University Press, 1930), passim.

27. Arnold L. Gesell, *The Mental Growth of the Preschool Child* (New York: Macmillan, 1925), passim; Gesell, *Infancy and Human Growth* (New York: Macmillan, 1928), passim; Sears, *Your Ancients Revisited. A History of Child Development,* pp. 52–55.

28. Mandel Sherman and Irene Case Sherman, "Sensori-Motor Responses in Infants," *Journal of Comparative Psychology,* 1925, *5:* 53–68.

29. Harold E. Jones and Mary Cover Jones, "A Study of Fear," *Childhood Education,* 1928, *5:* 136–143.

30. Arthur I. Gates and Grace A. Taylor, "An Experimental Study of the Nature of Improvement Resulting from Practice in a Mental Function," *Journal of Educational Psychology,* 1925, *16:* 583–592; idem, "An Experimental Study of the Nature of Improvement Resulting from Practice in a Motor Function," *Journal of Educational Psychology,* 1926, *17:* 226–236.

31. Florence L. Goodenough and C. R. Brian, "Certain Factors Underlying the Acquisition of Motor Skills by Preschool Children," *Journal of Experimental Psychology,* 1929, *12:* 127–155.

32. Arnold Gesell and Helen Thompson, "Learning and Growth: An Experimental Study by the Method of Co-Twin Control," *Genetic Psychology Monographs,* 1929, *6:* 1–124.

33. Josephine R. Hilgard, "The Effect of Early and Delayed Practice on Memory and Motor Performances Studied by the Method of Co-Twin Control," *Genetic Psychology Monographs,* 1933, *14:* 493–567; see also idem, "Learning and Maturation in Preschool Children," *Journal of Genetic Psychology,* 1932, *41:* 36–56.

34. See, for example: J. A. Hicks, "The Acquisition of Motor Skill in Young Children," *Child Development,* 1930, *1:* 90–105; Hicks and E. W. Ralph, "The Effects of Practice in Tracing the Porteus Diamond Maze," *Child Development,* 1931, *2:* 156–158; Mary Shirley, "A Motor Sequence Favors the Maturation Theory," *Psychological Bulletin,* 1931, *27:* 204–205; A. T. Jersild and S. F. Beinstock, "The Influence of Training on the Vocal Ability of Three-Year-Old Children," *Child Development,* 1931, *2:* 272–291; Jersild, *Training and Growth in the Development of Children* (New York: Columbia University, Teachers College, Bureau of Publications, 1932).

35. Nancy Bayley, "Mental Growth during the First Three Years: A Developmental Study of Sixty-One Children by Repeated Tests," *Genetic Psychology Monographs,* 1933, 14: 1–92. For further studies from the California Institute in the 1930s which furthered this line of thinking, see, for instance: Nancy Bayley and Harold E. Jones, "Environmental Correlates of Mental and Motor Development: A Cumulative Study from Infancy to Six Years," *Child Development,* 1937, *8:* 329–341; and Marjorie Honzik, "The Constancy of Mental-Test Performance during the Preschool Period," *Journal of Genetic Psychology,* 938, *52:* 285–302. By the 1950s the same Berkeley scientists, working with the same longitudinal projects (and subjects), had completely changed their point of view. See, for example, Nancy Bayley, "Consistency and Variability in the Growth From Birth to 18 Years," *Journal of Genetic Psychology,* 1949, *75:* 165–196.

36. See, for example: Committee on Child Development, Division of Anthropology and Psychology, National Research Council, *Conference on Research in Child Development. Gramatan Hotel, Bronxville, New York, October 23 to 25, 1925* (Washington, D.C.: National Research Council, 1925), passim; idem, *Second Conference on Research in Child Development, Washington, D.C., May 5–7, 1927* (Washington, D.C.: National Research Council, 1927), passim; idem, *Third Conference on Research in Child Development, University of Toronto, Toronto, Canada, May 2–4, 1929* (Washington, D.C.: National Research Council, 1929), passim; idem, *Fourth Conference on Research in Child Development, The University of Chi-*

cago, Chicago, Illinois, June 22–24, 1933 (Washington, D.C.: National Research Council, 1933), passim.

37. Carl Murchison, ed., *A Handbook of Child Psychology* (Worcester: Clark University Press, 1931), passim, quotes on p. ix.

38. Carl Murchison, ed., *A Handbook of Child Psychology,* 2nd ed., revised (Worcester: Clark University Press, 1933), passim. My special thanks go to Michael M. Sokal who made determined inquiries to ascertain if there were files among the Clark University Press archives that pertained to the two volumes, and who also confirmed that Murchison's own papers were lost in a fire in the early 1960s.

39. On the emergence of the Society for Research in Child Development, see, for example, Committee on Child Development, Division of Anthropology and Psychology, National Research Council, Committee Meeting Minutes, 1924–1933 folder, plus folders, Outline File, General File (discussion), Solution File, Archives, National Research Council, Washington, D.C.; see also Sears, *Your Ancients Revisited. A History of Child Development,* pp. 18–19. See also Committee on Child Development, Division of Anthropology and Psychology, National Research Council, *Proceedings of the 1st Biennial Meeting, Society for Research in Child Development, National Research Council, Washington, D.C., November 3–4, 1933* (Washington, D.C.: National Research Council, 1933), passim.

40. Examples of Gesell's many publications in these years are Arnold L. Gesell, et al., *An Atlas of Infant Behavior. A Systematic Delineation of the Forms and Early Growth of Human Behavior Patterns,* 2 vols. (New Haven: Yale University Press, 1934); Arnold L. Gesell, et al., *The First Five Years of Life. A Guide to the Study of the Preschool Child* (New York: Harper and Brothers, 1940); Gesell, *The Embryology of Behavior. The Beginnings of the Human Mind* (New York: Harper and Row, 1945) in collaboration with Catherine S. Amatruda; Gesell and Amatruda, *Developmental Diagnosis, Normal and Abnormal Child Development, Clinical Methods, and Pediatric Applications,* 2nd ed., revised and enlarged (New York: Paul B. Hoeber, 1947 [1941]). Gesell explicitly praised Coghill's work in Gesell, "Scientific Approaches to the Study of the Human Mind," *Science,* 1938, *88:* 225–230.

41. Leonard Carmichael, ed., *Manual of Child Psychology* (New York: Wiley, 1946), passim.

Gregg Mitman

Richard W. Burkhardt, Jr.

7

Struggling for Identity: The Study of Animal Behavior in America, 1930–1945

In an article in *Auk* in 1916, the young English biologist Julian S. Huxley, who had spent three years in the United States as a professor of biology at Rice Institute, perceptively remarked on some of the major differences that characterized animal behavior studies in his day. He observed that whereas trained biologists disdained the amateur's methods and his "failure to see principles behind facts," amateurs in turn disliked—"often with justice"—the professional's "dogmatism and his reliance on purely laboratory methods." At the same time, the major groups of scientists interested in behavioral issues—evolutionists, physiologists, and psychologists—each had their own conceptions of what constituted a satisfactory answer to such a question as "why does such-and-such a species of bird perform such-and-such an action?" The problem with their answers, Huxley allowed, was not that they were incorrect, for they were not, but rather that they did not take "a sufficiently broad point of view." They failed "to distinguish between ultimate cause, immediate cause, and mere necessary machinery." Huxley called for a collaboration between amateurs and professionals and for "a sufficiently broad point of view" that would allow one to see beyond the parochial boundaries of one's own discipline.[1]

Huxley returned to England in 1916 and never provided behavior studies with the sort of synthetic overview he sought to contribute to the study of evolution. In the United States on the eve of the First World War, the enterprise offering the most promise with regard to combining the efforts of biologists and psychologists was Robert Yerkes's *Journal of Animal Behavior*. But Yerkes's journal was itself an early casualty of the war, discontinuing publication when Yerkes entered the Army. The journal's self-identified successor, *The Journal of Comparative Psychology,* soon became an organ of little interest to zoologists. Over the next three decades, American zoologists and comparative psychologists were aware of each other's work, but by and large they pursued distinct research agendas. Without an umbrella professional organization, and with zoologists and psychologists publishing behavior studies in diverse journals, there were no bridging mechanisms to establish a unified field.[2]

Between the end of the First World War and the beginning of the Second, three institutional settings were particularly important for the promotion of biological studies of behavior in the United States. These were the Department of Zoology at the University of Chicago, where Warder Clyde Allee developed an approach rooted in animal ecology; the Department of Experimental Biology at the American Museum of Natural History, where Gladwyn Kingsley Noble was intent on uncovering the neural basis of social behavior in the vertebrates; and the American ornithological societies, especially the American Ornithologists' Union, where amateurs and professionals had the opportunity to interact and where naturalistic field studies received strong encouragement.

Laboratory and field studies of animal behavior were sharply divided in the interwar period, a division often drawn along professional versus amateur lines. The laboratory had been the focal point for American biological researches—including researches on behavior—since early in the century. Intimately tied to notions of professional identity and legitimacy, the laboratory became the professional biologist's trademark. It also served as the appropriate setting for analyzing physiological mechanisms influencing behavior, an approach encouraged by the availability of funds from the National Research Council Committee for Research in Problems of Sex to undertake laboratory studies of neural, hormonal, and psychological factors regulating sexual behavior and social interactions. Although Allee and Noble differed in their approaches to behavior, one explaining behavior in ecological terms and the other explaining it in evolutionary terms (what Huxley referred to as immediate versus ultimate causation), they both chose the laboratory as the location for their research efforts.

In contrast, the development of what Huxley had called "the science of the behavior of birds in their natural environment" was pursued as much by amateurs as by professionals. Lacking resources of the same magnitude as those available to at least some academic biologists, but at the same time independent of some of the constraints upon the professional, amateurs were in a position to make valuable contributions to behavior studies, particularly with regard to behavior of animals in the wild. It was an amateur, Margaret Morse Nice, who not only pioneered in the study of behavioral ecology but also provided the primary source of contact in the mid-1930s between American researchers and the nascent continental tradition of ethology. When the New York Zoological Society in 1947 established a Committee for the Study of Animal Societies under Natural Conditions, it was not only reacting against the general hegemony of laboratory work in American academic zoology but also responding favorably to the kind of field studies that Nice and others had been successfully pursuing. These studies opened the door for the reception of continental ethology and for a rebirth of field research by a new generation of American students of animal behavior.

Our aim in this essay is to analyze the importance of individual initiatives, disciplinary affiliations, institutional settings, and contingent events in shaping animal behavior studies in American zoology at the time when conceptual, methodological, and institutional dimensions of such studies remained uncertain and in flux. Psychologists and physiologists as well as zoologists—and amateurs and field naturalists as well as laboratory professionals—all believed they had something special to offer in understanding animal behavior. The professional identity of American zoologists contributing to the study of behavior came first and foremost from the disciplines within which they were trained. These differences in disciplinary training were reflected in aims and methods that different investigators espoused. With disparities among professionals as well as between professionals and amateurs, Huxley's call for a sufficiently broad approach to the study of animal behavior went unanswered.

The Chicago Tradition

Diverse approaches to the study of animal behavior taken by professional biologists at the turn of the century are well exemplified by the work of Charles Otis Whitman and Jacques Loeb. Both Whitman and Loeb had been at the University of Chicago since it opened in 1892. Whitman was the founder of the Department of Biology (later the Department of Zoology) at Chicago, and Loeb was Whitman's choice as professor of physiology. Both regarded animal behavior as a subject rich in possibilities, a subject indeed central to the properly conceived study of animal *life*. But where Whitman was primarily concerned with questions of ultimate causation, promoting a broadly conceived "experimental natural history" in which the organism's evolutionary history and heredity were of central concern, Loeb was basically interested in questions of immediate causation and in the manipulation and control of behavior in the laboratory. Whitman addressed issues of the relations between instinct and intelligence, the means by which behavior has evolved, and the value of instinctive behavior patterns in reconstructing phylogenies. He not surprisingly took exception to Loeb's studies of tropisms, where, as Whitman complained, "the whole course of evolution drops out of sight altogether." Loeb's tropism studies, nonetheless, had a powerful appeal for a new generation of investigators who saw experimentation in the laboratory as the forefront of biological research and the key to their own careers. Although Herbert Spencer Jennings's experimental studies of the behavior of the simpler invertebrates, focusing on internal physiological states and individual variability, showed that Loeb's bold interpretations of behavior could not accommodate the detailed analysis of the actions of a variety of different organisms, the objective and experimental methods that Loeb and Jennings promoted had a marked influence on contemporary biological research. Jennings's work may have led individual investigators to despair of explaining the behavior of animals in terms of mechanical responses to external stimuli, but the physiological approach he and Loeb advocated found a place in the new field of animal ecology, particularly as this field was developed at the University of Chicago from 1910 to 1950.[3]

Victor Ernest Shelford was the central figure in the development of a program of animal ecology at Chicago during the first fifteen years of the twentieth century. He began his career at Chicago as a student, entering the zoology department as an undergraduate in 1901 and completing a doctoral dissertation on the "Life Histories and Larval Habits of the Tiger Beetles" six years later.[4] Interested in determining the causal factors governing animal distributions, he embraced the physiological approach to behavior and morphology advocated by Loeb, Jennings, and his adviser, Charles Manning Child. Physiological animal geography would, as Shelford conceived it, study functional responses to arrive at an understanding of the distribution of organisms within the environment. He looked to the experimental science of physiology for the field's legitimization, just as botanists in the late 1890s emphasized the physiological basis of plant ecology research.[5]

In the same way that plant ecologists used form as an index of environmental conditions, Shelford maintained, so too could animal ecologists use behavior as a diagnostic tool for understanding relations between organism and environment. "The behavior and general mode of life of animals," he explained, "are the superficial equivalent of the structural phenomena in the vegetative parts of plants. Behavior and vegetative structure are convenient indices of physiological conditions within the organism."[6] He coined the term "mores" to refer to behavioral and physiological characteristics of a group of organisms as distinct from morphological characteristics of a species. In contrast to taxonomists, who classified animals according to morphological characters rather than activity or behavior, he sought a method of classification that related taxonomically diverse groups on the basis of their physiological response to the environment. Studying experimentally the physiological life histories of individual plants and animals thus became the first step in determining the factors governing ecological distribution.

After receiving his doctorate in 1907, Shelford remained at Chicago as an associate and then instructor on the zoology faculty. He directed his attention toward establishing a program of research in behavior and ecology, instituting a whole sequence of courses including field zoology, animal ecology, geographic zoology, animal behavior, and a graduate seminar on topics in ecology conducted in conjunction with the botany department. The courses at Chicago were not the only place where behavior and animal ecology were conjoined; behavior and ecology also represented a combined specialized grouping at the 1912 meetings of the American Society of Zoologists.

Shelford directed a number of dissertations at Chicago that attempted to correlate a species' physiological life history with its distribution in the wild. In each case, the researcher analyzed behavioral responses in the laboratory to environmental factors deemed important in the organism's habitat, determining which environmental conditions governed the animal's distribution based on the behavioral response. For example, Allee, a graduate student of Shelford's who entered the Department of Zoology at Chicago in 1908, found that differences in the behavior of pond and stream isopods to water currents, and consequently differences in their distributional patterns, were dependent on variations in oxygen and carbon dioxide concentrations found in their respective habitats.[7] Similarly, in examining the reactions of fishes to gradients of dissolved gases, Shelford and Allee suggested that one could predict whether certain species of fish would be found in a particular location depending on the carbon dioxide content in the water. Often, the influence of breeding behavior on distributional patterns was also emphasized.[8]

Shelford left Chicago in 1914 for the University of Illinois, but despite his absence, the behavioral and laboratory orientation of Chicago animal ecology flourished. In 1915 Morris M. Wells, a former student of Shelford, was hired to replace the vacant position left by his adviser. Although Wells continued to teach the ecology course sequence, he did not actively pursue further research and in 1919 he resigned. Wells's replacement, William J. Crozier, himself a researcher in animal behavior, left after one year. In 1921 Allee, then at Lake Forest College, accepted an offer to return to the University of Chicago as a faculty member in the Department of Zoology.[9] Frank R. Lillie, chairman of the department, described the position to Allee as a "succession to the one originally held by Dr. Shelford, and . . . includes more particularly the subjects of Ecology and Animal Behavior."[10]

From 1921 until his retirement in 1950, Allee continued to offer the course on animal behavior. The course focused almost exclusively on tropisms and organismic responses to environmental factors such as gravity, light, temperature, and moisture, with some attention given to the topics of learned versus instinctive behavior and the evolution of the nervous system. In emphasizing the interaction between the organism and its abiotic environment, however, the course neglected another important aspect of the organism-environment relation: the influence of organisms on one another.

By the 1920s, ecological research in the United States and at Chicago was proceeding along two rather independent lines. Under the heading of individual ecology (also referred to as autecology), interest centered on the individual and its relation to the environment. Here the focus was on physiology and behavior, and the laboratory served as the organism's milieu. In contrast, community ecology (also referred to as synecology) was pursued in the field. Research was largely descriptive, and considerable attention was placed on the problem of succession. Allee's early career illustrates this dichotomy in animal ecology research. He continued to work on the ecological physiology of individual organisms in the laboratory and also spent time at Woods Hole or in such distant places as Barro Colorado Island in the Canal Zone pursuing community studies.[11]

In 1924 Allee added a graduate seminar entitled "Animal Aggregations" to the department's curriculum. The course centered on "a study of the different types of social organization and relations of animals" and "their origin and effect upon the individual and the species."[12] The addition of this course reflected a new avenue of research initiated by Allee to bring community analysis within the laboratory domain. Little was known about the nature of biotic interactions that structured animal communities. Allee's work on animal aggregations, on the "physiological effects of crowding upon the individuals composing the crowd," was an attempt to unravel the significance of "inter-mores" interactions, of the physiological effect of biotic relationships.[13] Through this study, Allee believed one could gain insights into the problems of community integration, the importance of biotic relations, and the origins of sociality. At the aggregation level, the same forces that operated within the community were present, but they acted in a much simpler and more accessible way.

Allee's early research compared the effects of grouped versus isolated organisms on metabolic rate. By holding all physical factors constant except for the number of individuals, Allee analyzed the effect of group membership on individual physiology. In land isopods, for instance, isolated individuals exhibited greater water loss over long periods than did grouped individuals. Similarly, during periods of starvation, bunches of brittle stars outsurvived their isolated kin. Hence individuals within a group were less susceptible to changing environmental conditions than when alone. During the 1930s and 1940s Allee and his students continued to gather evidence on the beneficial effects of crowding. Experiments demonstrated mass protection against toxic reagents, increased growth of individuals in biologically conditioned water, and higher reproductive rates at intermediate densities.[14]

Through this research, Allee believed he had found experimental evidence indicating the existence of an integrating factor in animal communities more subtle than traditional food relationships, a factor that he identified as nonconscious cooperation, or the "auto-protective value of the community."[15]

Allee's research had implications not only for animal ecology but for a general theory of sociality as well. During the interwar years a number of biologists conducted research on animal societies with the hope of erecting a comparative sociology. Of all the biological disciplines, ecology contributed the most to this endeavor. Plant and animal communities were simply one form of association, and ecology texts such as Josias Braun-Blanquet's *Plant Sociology* (1928), a text that Allee used in his animal aggregation course, were laden with sociological metaphors. Allee's own studies were pursued within this general context. He had, as early as 1923, laid the interpretive framework for this research. For Allee, a continuity existed between the loosely integrated associations of the ecologist and true societies. Sociality was not to be derived from the family but instead arose from "the consociation of individuals for cooperative purposes."[16] Human society was but an extension of the associations that existed throughout the animal kingdom—associations that arose as a consequence of the mutual benefit that group life provides in the individual's struggle with its environment. Rejecting the notion of a social instinct, Allee argued that the "first step toward the development of social life is the appearance of a physiological tolerance for the presence of other animals in a limited area where they have collected as the result of their individual reactions to the conditions obtaining within their environment."[17] Once formed, the aggregation was maintained through benefits acquired in group life. An active participant in the liberal pacifist movement during World War I and a member of the American Friends Service Committee throughout World War II, Allee offered a theory of sociality that was interdependent with his own hopes to restructure human society on an egalitarian and cooperative basis.[18]

Allee's analysis of social behavior centered not on phylogeny, not on similarities due to common ancestry, but on the similarities of behavioral responses in unrelated organisms to the more immediate effects of environment. His interest in physiology and environment and his lack of enthusiasm for hereditary explanations of behavior fit well within the context of the department. Under the chairmanship of Lillie, zoology at Chicago centered on developmental research, especially as it related to the biology of sex and the physiology of reproduction. In general, the faculty had a common view of development, interpreted as an interactive process between organism and environment that was constrained but not determined by underlying genes. The journal *Physiological Zoology,* founded by Child in 1928, emphasized the department's environmental and physiological outlook, and the journal became an important outlet for the publication of behavioral physiology research when Allee took over its editorship in 1935.[19]

Under Lillie's reign, however, ecology and behavior always remained something of a marginal field. When the chairmanship of the department became vacant in 1934, Allee was ambitious for the appointment, since ecology "had not received quite the recognition that its followers deserve." He felt compelled to work toward improving the professional status of ecology, and he saw the chairmanship as a possible vehicle for increasing the visibility of this field. But Lillie, who was then dean of the Division of Biological Sciences, was eager to see the embryological tradition live on. In the 1930s a series of operations for a benign spinal tumor left Allee paralyzed from the waist down. Lillie dissuaded him from taking the chairmanship, citing Allee's physical disability as the main reason. In his conversation with Allee, however, Lillie also "suggested vaguely something about the developmental field remaining the center of the department's program."[20] Not surprisingly, Carl R. Moore, Lillie's graduate student and then colleague, assumed the chairmanship. In addition, Allee had great difficulty obtaining research funds after 1934 because of agreements that Lillie made with the Rockefeller Foundation and the National Research Council Committee for Research in Problems of Sex (NRC-CRPS).

Lillie, along with Fred C. Koch in the Department of Physiological Chemistry and Pharmacology, received approximately $24,000 per year between 1928 and 1934 from the NRC-CRPS for work in endocrinology and the physiology of sex. In addition, the Rockefeller Foundation provided $30,000 per year from 1929 to 1934 to the Division of Biological Sciences at Chicago.[21] In 1934, however, the Rockefeller Foundation combined the NRC funding for sex research at Chicago with the general biological fund into a single grant of $50,000 per year, with the understanding that money should not be given for research that fell outside the program interests of the foundation. This included, most notably, ecological research under Allee. His general biological fund grant (on the order of $3,000 per year) was transferred to support Sewall Wright's work in genetics. To counterbalance Allee's loss of support, the department did provide him with an equivalent amount of money from separate departmental funds that were previously used to support Wright. Although not immediately apparent, the new funding arrangements made by Lillie, the Rockefeller Foundation, and the NRC-CRPS jeopardized Allee's behavior research, only in more indirect ways.[22]

During this period Allee had become interested in studying social organization in the vertebrates and, specifically, dominance-subordinance hierarchies in birds. Spurred on by the claims of the Norwegian biologist T. Schjelderup-Ebbe that despotism was the fundamental principle underlying the social organization of birds and stimulated by the research and availability of funds in the department for work on sex hormones, Allee proposed injecting hormones into birds of known social rank to examine the effects on group organization. His interest in flock organization was merely another attempt to understand physiological mechanisms responsible for group integration. He applied to the NRC-CRPS in 1935 for support but was forced to withdraw his application. Under the terms set by Lillie, the Rockefeller Foundation, and NRC-CRPS, individual faculty from Chicago were not allowed to submit applications directly to the Rockefeller Foundation or NRC-CRPS for research in the biology of sex, since this was already covered in Chicago's general biology fund.[23] Unable to get support from the NRC-CRPS, Allee's research on hormones and behavior was delayed for over two years. In contrast, G. K. Noble at the American Museum of Natural History had little difficulty securing funds from the NRC committee from 1935 to 1940 for similar experiments.

Allee eventually managed to secure a modest $4,000 from the NRC Committee for Research in Endocrinology for his hormonal research between 1937 and 1939. He and his students, including Hurst Shoemaker, Nicholas Collias, Catherine Lutherman, and Elizabeth Beeman, systematically explored the effects of hormones (testosterone, estrogen, epinephrine, and thyroxine) on the social rank of individual hens. When, for example, a low-ranking individual in a flock of white leghorns was injected with testosterone proprionate over an extended period, the injected individual won a greater portion of staged contacts and rose in social status, eventually occupying the dominant position. Further research also indicated the importance of conditioning and learning in modifying individual aggressiveness.[24]

Allee was primarily interested in how behavior served as an integrating mechanism in animal societies and less in its evolutionary origins. In later years, however, he did develop an evolutionary account for his theory of sociality and the significance of group organization, a theory strongly influenced by the writings of his colleagues Alfred Emerson and Sewall Wright. Allee argued that because of the mutual benefit that group life provided in the individual's struggle with its environment, groups that displayed greater cooperation and integration among individual members would be continually selected for. "Evolution," Allee wrote, "even the evolution of individual organisms is a process which takes place on the group level only."[25] The population as a unit of selection was a theme characteristic of Chicago ecology, and one invoked by Allee to explain the origins of social life.

Animal ecology at Chicago owed much to the physiological tradition of behavior research. Animal behavior helped bring the authority of experimentation to ecology during a period when ecology was itself seeking professional legitimation. In the process, however, the study of behavior became in part subsumed under the subdiscipline that it initially helped foster. And although the emphasis on the laboratory and experimental analysis may have brought greater prestige, it was not without cost. When Nikolaas Tinbergen published *Social Behavior in Animals* in 1953, he criticized the American literature for overemphasizing the importance of peck-order as a principle of social organization in the vertebrates. Tinbergen's comments point to the laboratory bias of Chicago behavioral studies, for it was in the laboratory that the peck-order received its greatest verification.[26] Only a handful of studies existed in the early 1940s that documented the presence of dominance hierarchies under natural conditions.[27] In contrast, numerous studies undertaken primarily by ornithologists and amateurs had documented territorial behavior in the field.[28]

A number of Allee's students were influential in founding the Section on Animal Behavior and Sociobiology of the Ecological Society of America and the American Society of Zoologists in 1956.[29] Yet this new postwar generation of animal behavior researchers left the confines of the laboratory for the field, bringing experimental techniques such as time-motion studies and sound spectrographs into the wild, studying the organism's behavior within its natural environment. Now that the disciplinary status of ecology was ensured, the emphasis on physiological analysis and experimentation that was so much a part of Allee's early training was no longer a central issue. And the emphasis on sociality and group structure that marked the Chicago program also faded into the background as behavior was reduced to either the instincts and social releasers (see p. 178) of the ethologist or the language of selfish genes, individual reproductive success, and reciprocal altruism that has come to symbolize modern sociobiology.

The American Museum of Natural History

The lack of institutional support for Allee's behavior work stands in marked contrast to the experiences of Gladwyn Kingsley Noble (1894–1940) at the American Museum of Natural History. Noble, son of the noted publisher Gilbert Clifford Noble, had developed a fascination for studying the life histories of reptiles and birds while an undergraduate at Harvard.[30] He received his A.M. degree from Harvard under Thomas Barbour in 1918 and went on to obtain his Ph.D. from Columbia University in 1922. His dissertation on "The phylogeny of the Salietia" reflected his lifelong interest in herpetology and systematics.[31] Through William K. Gregory, Noble's graduate adviser, professor of vertebrate paleontology and curator of the Department of Comparative Anatomy at the American Museum, Noble also developed an appreciation of functional morphology and microscopical anatomy in analyzing phylogenetic relationships.[32]

The close ties between Columbia and the American Museum led to Noble's appointment as curator of the Department of Herpetology at the latter institution in 1924. But Noble was not content with a career devoted entirely to systematics and natural history when the cutting edge of biology lay in the experimental disciplines of neurology, physiology, and endocrinology. In the spring of 1928 Noble was offered the position of professor of experimental zoology at Columbia to replace Thomas Hunt Morgan, who had accepted a post at the California Institute of Technology. Henry Fairfield Osborn, president of the American Museum, advised Noble to decline the offer, indicating that Noble could not expect "any extensive plan of cooperation with the Museum in developing the broad and comprehensive research program expected at Columbia." Taking full advantage of the situation, Noble agreed to stay if the museum would give full support for research work in experimental biology and would supply him with an annual fund of $4,000 for experimental research and a salary increase of $1,000 per year. The trustees agreed, and on May 7, 1928, the museum changed the name of the Department of Herpetology to the Department of Herpetology and Experimental Biology, with Noble serving as curator.[33]

Two months later, when Cornell University Medical School offered Noble a full professorship of microscopical anatomy, Noble raised the stakes even higher. The museum, he indicated, would have to meet a rather staggering list of demands in order to keep him. The most significant included: (1) an increase in the annual budget of the department by $10,000 above its current annual budget of $17,000 with a guarantee of five years; (2) additional space for a laboratory within the existing building occupied by the department; (3) half of the fifth and six floors of the new African Hall for a laboratory of experimental biology, with Noble given full license as to the arrangement and installation of the labs and complete control over the nature of biological research; and (4) permission for Noble to devote half his time to experimental biological research. Perhaps realizing the prestige and attention that such a laboratory might bring, the museum acquiesced to Noble's demands.[34]

Noble's intention in creating a Department of Experimental Biology was to establish a center where natural history and experimental biology would meet. He regarded the "museum with its corps of field men and its many specialists on the forms of life" as "the logical place for applying the experimental method to the problem of the naturalist."[35] Over the next five years, much of Noble's experimental research was devoted to the effects of endocrine secretions on processes such as reproduction, tooth formation, molt, and brooding reactions in the amphibia. In 1930 Noble was able to induce egg-laying in a species of salamander through transplants of the anterior portion of the pituitary gland. He regarded this discovery of fundamental significance, for now the naturalist could secure information about the breeding habits of a wild species and obtain embryological material at will.[36]

In May of 1934 the city of New York completed Noble's laboratory of experimental biology. It occupied the entire sixth floor and roof of the African Hall and included, among other things, an aquarium room, three greenhouses, an animal house, histology laboratory, and physiology laboratory, at an expense of $78,920.[37] But because of the financial constraints caused by the Great Depression, the museum could no longer maintain its previous level of support. The department's budget for 1934 was cut by $7,000 and was slashed another $7,000 in 1935.[38] To adjust for the loss of his research and clerical staff, Noble managed to secure help from the Works Progress Administration. In 1934 he started with seventeen WPA people working in his laboratory; by 1937 this number had escalated to sixty-five. Not only were these workers involved in the preparation of exhibits and the maintenance of the aquarium rooms and laboratories, but a number of individuals also worked as research assistants. In addition, Noble had a staff of at least seven people responsible for translating biological articles from foreign journals, analyzing the literature dealing with the morphology, physiology, and habits of reptiles, and collecting literature on the courtship and sexual behavior of animals.[39]

Facing budgetary reductions, Noble decided to limit his research to the "physiology and psychology of reproduction in the lower vertebrates" even though the laboratory was originally intended to "consider many problems on the borderline between natural history and biology."[40] He regarded his research on reproductive physiology and behavior as the most significant of his laboratory studies. It was also an area that was being actively supported by the National Research Council Committee for Research in Problems of Sex, a fact that surely swayed Noble's interests. Having spent the summer quarter at the University of Chicago in 1931 with the hope of borrowing "many ideas and techniques" for his own laboratory, Noble left Chicago impressed by Lillie's large-scale enterprise in reproductive physiology.[41] To be sure, Noble's earlier work on the biology of the amphibia demonstrated an interest in the physiological and psychological problems associated with reproductive behavior. But the NRC-CRPS certainly helped reinforce Noble's interest. An adept entrepreneur, he was careful never to distance himself from the public eye, and sex was surely a topic of public interest. As Douglas Burden, a museum trustee, film producer, and close friend remarked to Noble: "What excitement you will evoke if the substance of your new hormone not only brings immature animals to sexual activity but increases sexual activity among the aged. . . . Pull that trick and we will never have any difficulty in raising funds for your research."[42] Taking Burden's comments to heart, Noble went public, publishing an article in the *New York World-Telegram* entitled "Recent Advances in Our Knowledge of Sex."[43]

In *The Biology of the Amphibia*, published in 1931, one finds much of the conceptual background for Noble's later behavioral studies already in place. Noble, unlike Allee, viewed behavior within the context of phylogenetic relationships and instincts, analyzing similarities in behavior on the basis of common ancestry rather than environmental relations. His perspective was that of systematist rather than ecologist. In the chapter "Instincts and Intelligence," Noble devoted considerable discussion to the evolution of courtship behavior in salamanders. He detailed how courtship patterns of various families of salamanders were all modifications of the pattern found in the most primitive group. This courtship behavior was, according to Noble, a clear example of instinctive behavior. In Noble's definition, instinct signified "a normal response of inherited patterns of neurons to certain sensory stimuli." Overlaid on these instinctive patterns were factors such as hormones, age, and nutrition that could affect behavior by "modifying the synaptic resistances." But the fundamental origins of behavior lay in the functional organization of the nervous system. Remarking that "the centers in the central nervous system controlling this [courtship] or any other instinct have never been determined," Noble hinted at a research agenda that would occupy the rest of his career.[44]

Beginning in 1935, Noble's research was continually funded by the Josiah Macy, Jr. Foundation and the NRC- CRPS.[45] The overall direction of Noble's program was to utilize the techniques of endocrinology and neural surgery to establish a detailed picture of the mechanisms responsible for social behavior in the evolution of vertebrates. By analyzing the social behavior of fishes, amphibians, reptiles, birds, and finally mammals, Noble hoped to ascertain how far phylogenetic changes in neural structure had led to differences in social behavior patterns. Like Allee, he was laying the basis for a comparative sociology, but the underlying foundation was radically different. For Noble, behavior was to be understood in neurophysiological structures and processes ingrained in the individual organism as a consequence of its phylogenetic past.

Two examples taken from Noble's studies on sexual behavior in fishes and birds will help illustrate the extent of his research program. Noble always began by first establishing a detailed account of the social behavior of a given organism within its natural environment. In his study on the social behavior of the jewel fish, for example, he provided an exhaustive description of the schooling behavior, sexual behavior, brooding behavior, and the factors important in territory and sex recognition of this class of cichlids.[46] Once he had established a firm informational base on the behavior of a species in the wild, he would then determine to what extent, if any, these patterns were modified by laboratory conditions.

Noble's next step was to isolate the hormonal and neural factors responsible for these patterns. Beginning with the most proximate cause, he observed the sexual and brooding behavior in gonadectomized fish. In three families of fishes, he found that castration in males had no influence on the breeding or brooding cycle and, hence, that these behaviors were not directly under the control of a testicular hormone. Seeking a more rudimentary cause, Noble next made a series of ablations in the forebrain and found that the breeding cycle was seriously affected. From these studies, he concluded that in fish the dorsal portion of the corpus striatum was "necessary for successful integration of the mating and brooding reactions" and was thus the site of the instinctive reproductive patterns.[47]

Noble used the same methodology in his studies on the social behavior of birds. His published account of the black-crowned night heron contains a detailed description of the bird's life history, including a discussion of pair-formation, dominance, the significance of territoriality, and other details of its sexual and brooding behavior.[48] Noble died, however, before the neurophysiological portion of this study was published. In this unpublished research he had analyzed the influence of androgens and estrogen on the night heron's sexual behavior, and through a series of forebrain lesions he had determined the neural centers involved.[49] The cortex, according to Noble, played an important function in territorial behavior and egg-laying, whereas the corpus striatum was "essential for the integrating of more complex features of the social behavior."[50]

In January 1938 Frank Beach was hired as assistant curator in the Department of Experimental Biology.[51] Noble had at last found someone to investigate the neural basis of social behavior in mammals, the one group in the vertebrate series that his laboratory had not yet explored. Beach had been a student in Karl Lashley's laboratory at Harvard and was interested in the neural mechanisms governing sexual behavior in the rat. With Beach's work, the comparative sequence was complete. Noble noted with confidence that "there has been a shift of the brain centers necessary for the adjustment between two or more individuals of a social group, from the corpus striatum of fish and birds to the cortex of mammals." "With the elaboration of the cortex in primates," he continued, "there followed other improvements in social behavior. Tradition became more important than in lower forms and insight into the benefits of cooperation formed an important advance. Many of the old components of social behavior, such as that of dominance, were greatly modified. It is these improved components of social integration that form the basis of human society."[52] Through the study of social behavior in the lower vertebrates, Noble sought a comparative basis for understanding the origins of human sociality, a comparison centered on changes in the neurophysiological structure of the brain.

Because Noble's analysis depended on observations of the social behavior of organisms in their natural environment, he relied heavily on the accounts of field naturalists and amateur observers. The museum, with its network of collectors and its high public visibility, was an institution that served Noble's needs quite well. "Our unique position," Noble perceptively wrote, "is assisting local field naturalists in an analytical study of animal behavior. Some of these students are professional ornithologists, others are graduate students, and still others are business men who spend their spare time in the field. This is a service which no university laboratory is equipped to render."[53] Indeed, Noble established an impressive network of field observers that could supply him with needed information. The Huyck Preserve in Albany, New York, the Conservation Department of New York state, the Kent Island preserve off the coast of Maine, Marineland in Florida, and numerous individual ornithologists supplied Noble with both material and field notes necessary for developing a comparative phylogenetic study of behavior. His laboratory also helped other investigators in the field. His WPA staff assembled an informational resource base of 10,000 abstracts of articles pertaining to field observations on the social and sexual behavior of animals. It was through Noble, furthermore, that a complete translation of Konrad Lorenz's seminal article of 1935, "Der Kumpan in der Umwelt des Vogels" ("Companions as Factors in the Bird's Environment"), was made in 1937 and distributed to various American biologists and psychologists.[54]

Lorenz in his "Kumpan" article began laying out the analytical and theoretical framework that was to serve as the foundation for continental ethology from the mid-1930s into the 1950s. The Austrian naturalist focused his attention on species-specific instinctive behavior patterns, which he had studied in the course of maintaining a large number of different bird species in a free-flying state. Maintaining that a sharp distinction had to be made between instinctive and learned behavior, he insisted on the special taxonomic value of instinctive behavior patterns (as had been done before him by both C. O. Whitman and the German zoologist Oscar Heinroth). He paid particular attention to "releasers" and "innate schemata," the former being special organs and instinctive behavior patterns that serve to elicit social responses in conspecifics and the latter being "ready-formed, species-specific functional layouts" that govern the instinctive responses of animals to releasers and other "sign stimuli." By interpreting special structures, conspicuous colors, and species-specific display behavior as social releasers, Lorenz succeeded in making sense of a wealth of biological data that previously had been accounted for only in part by Darwin's theory of sexual selection.[55]

Noble did not champion Lorenz's work. Although he saw to it that Lorenz's 1935 paper was made available in translation to American biologists and psychologists, and although in 1939 he offered Nikolaas Tinbergen, Lorenz's Dutch coworker, the position of resident naturalist at the Huyck Preserve, Noble had reservations about the ideas of the continental ethologists. He found fault with Lorenz's idea of releasers, and he did not think Lorenz had distinguished adequately between innate and learned behavior. Lorenz, in turn, found Noble's work to be flawed by its emphasis on dominance relations, which Lorenz regarded as a product of the artificial situations in which Noble studied animals. Each man had taken it upon himself to put the study of animal behavior on a broad and new foundation, and each was inclined to see the other as too ready to generalize on an inadequate basis of facts.[56]

Noble's aspirations to provide a broad foundation for animal behavior studies were brought to a sudden end in December of 1940 when he succumbed, at the age of forty-seven, to a streptococcus infection known as "Ludwig's quinsy." His Department of Experimental Biology was so much an expression of his own aspirations and energies that the director of the museum, Roy Chapman Andrews, assumed the department would be unable to continue without its founder. However, it was saved and renamed the Department of Animal Behavior, primarily through the efforts of Beach, who, after Noble's death, was the only permanent member of the scientific staff. Under the leadership of the psychologists Beach and later Lester Aronson and T. C. Schneirla, the department promoted an interpretation of animal behavior that stressed, among other things, the role of environmental factors in behavioral development. While Schneirla and Beach wrote important critiques of the way most American psychologists had studied animal behavior, Schneirla and Daniel Lehrman launched in the early 1950s a significant assault on some of the major assumptions of Lorenz. The positive reception in American in the 1930s and 1940s to the work of Lorenz and Tinbergen came not from the investigators of animal behavior at the American Museum but rather from ornithologists who were engaged in a powerful revival of field studies.[57]

Field Studies and the Contributions of M. M. Nice

Although in the mid-1930s there was no other research facility anywhere in the world—with the possible exception of Karl von Frisch's laboratory in Munich—that could compare to Noble's in its combination of resources and its commitment to the study of animal behavior, the fact remained that important studies of the behavior of animals in their natural environments could be accomplished independently of such institutional support. Lorenz and Tinbergen, for example, were operating at this time only on shoestring budgets. Although Lorenz had hopes of having a Kaiser Wilhelm Institute established for him, he was still living and working out of his father's home in Altenberg, Austria. In the meantime, Tinbergen, holding the junior and poorly paid post of assistant at the zoological laboratory at Leiden, conducted much of his research and directed the research of his students at the very frugally run summer research camp he had established at Hulshorst. What such field workers needed as much as anything was the time necessary to do their research, access to suitable animals and natural settings, and the institutional means of coordinating their activities and exchanging information and ideas. Insofar as most professional zoologists in this period were oriented toward laboratory or museum research, the way was open through field studies for these naturalists—and for amateurs as well—to make a major contribution to the development of behavior studies. This occurred primarily though not exclusively in the field of ornithology.[58]

Early in the century, amateurs had played perhaps the leading role in demonstrating the value to science of the careful study of common animals in their natural environments. In the United States, George and Elizabeth Peckham had conducted important researches on the instincts and habits of solitary wasps (having earlier done important work on the behavior of spiders). In Britain Edmund Selous had initiated the practice of recording in painstaking detail the behavior of birds in the wild, while Eliot Howard, patiently observing the British warblers, had brought the concept of territory to the fore. If a professional biologist like Julian Huxley believed he had an important role to play in making bird-watching scientifically respectable, he also recognized that amateurs enjoyed certain advantages that he lacked. His own major papers on bird courtship, written over a fourteen-year period, were the product of at most forty to fifty days of field work, carved out of his vacation time. As he complained to his friend Eliot Howard in 1922: "In term-time I get less than no time to think over bird problems—I am full up with pupils, committees, lectures, & other research."[59] In the United States the amateur who did the most to advance studies of bird behavior and to stimulate professionals and amateurs alike was Margaret Morse Nice.[60]

Margaret Morse grew up in Amherst, Massachusetts, with a keen love of nature and a particularly passionate interest in birds. She reports in her autobiography that as an undergraduate at Mount Holyoke College she enjoyed her teachers and friends and her opportunities to go exploring in the countryside, but she found little relation between her college zoology courses and the wild things she loved. Among other things, she "saw no future in laboratory zoology." After graduating from Mount Holyoke in 1906 and spending a year as a "daughter-at-home" with her family, she enrolled in graduate studies at Clark University, having been attracted there by Dr. Clifton F. Hodge's encouragement that one could indeed be a zoologist and study *living* animals. For two years she pursued graduate studies at Clark, conducting research on the food of the bobwhite. Her academic career in effect ended when she married Leonard Blaine Nice, a graduate student in physiology. With regard to not having gone on for a Ph.D., she later wrote: "Sometimes I rather regretted that I had not gone ahead and obtained this degree. . . . But no one had ever encouraged me to study for a doctor's degree; all the propaganda had been against it. My parents were more than happy to have me give up thoughts of a career and take up home-making."[61] Taking on the role of housewife and mother, she concentrated her scientific talents on observing the psychological development of her children and pursuing a variety of ornithological subjects. While her various accomplishments show how well she adapted to the locations where her husband's career took the family—to Norman, Oklahoma in 1913; to Columbus Ohio, in 1927; and to Chicago, Illinois in 1936—her story shows how gender-related cultural expectations provided major obstacles to her scientific talents and aspirations.

The course of Nice's career as an ornithologist was shaped not only by gender and the sorts of contingent events and personal talents that contribute to any career but also by a particular set of institutions that favored the interaction of amateur and professional ornithologists—the American ornithological societies. The two most important of these societies for Nice were the American Ornithologists' Union (AOU) and the Wilson Ornithological Club. Founded in the late nineteenth century, the AOU and the Wilson Ornithological Club (together with the Cooper Ornithological Club) were important institutions for the interaction of professional and amateur ornithologists. Ernst Mayr has indicated that "perhaps nowhere else in the world have ornithological societies played such an important role in stimulating ornithology as in North America."[62] Fieldwork in particular was stimulated in the early decades of the twentieth century by the advent of new technologies ranging from better means of transportation to improved photographic equipment and the new technique of keeping track of individual birds by placing distinctive bands on their legs.

Nice became committed to ornithological research in 1919 when, provoked by a proposed change in the Oklahoma state game laws, she undertook a study of the nesting times of mourning doves. She reported on her mourning dove work at the meeting of the AOU in Washington, D.C. the following year. At Washington she also learned of the Wilson Ornithological Club, centered in the Midwest, which she joined in 1921. At the 1922 meeting of the AOU in Chicago she met the accomplished amateur Althea Sherman, whose example provided her with both inspiration and instruction.[63] When her husband's career took the family from Oklahoma to Columbus, Ohio, in 1927, Nice deeply regretted having to leave the birds and bird haunts of Oklahoma that had become so familiar to her, but she found that the move had the advantage of bringing her much closer to the ornithological meetings.[64] In 1927 she went to the AOU meeting in Washington and to a joint meeting of the Wilson Club and the Inland Bird Banding Association in Cleveland. The following March she banded her first song sparrow.

The research on song sparrows that established Nice's scientific reputation reveals a fine interplay between the challenges of fieldwork on the one hand and the stimuli provided by interactions with additional investigators on the other. In her autobiography, she describes how witnessing a territory-establishment conflict between two song sparrow males in 1929 "sealed my fate for the next fourteen years. I was so fascinated by this glimpse behind the scenes with my Song Sparrows that I then and there determined to watch Uno [the first sparrow she named] for several hours every day, to find out the meaning of his notes and postures, in short, to discover exactly what he did and how he did it."[65] At the same time, her autobiography reveals how important the ornithological meetings were becoming for her.[66]

In the 1930s, as her work on song sparrows progressed, Nice became actively involved with many of the world's leading ornithologists, and she proceeded to become an important element in the international exchange of ornithological information and ideas. At the 1931 meeting of the AOU in Detroit, she met the young German ornithologist Ernst Mayr, who was working at the American Museum in the Department of Ornithology. Mayr set Nice reading the *Journal für ornithologie* immediately. He also put her in touch with leading German ornithologists, including Oscar Heinroth and Erwin Stresemann. She met both Heinroth and Stresemann in Germany in 1932 (having traveled to Europe with her husband for the Fourteenth International Physiological Congress), and Stresemann encouraged her to send her work on the song sparrow to him for possible publication in the *Journal für Ornithologie*. Her first major account of her song sparrow work appeared in German in Stresemann's journal as a two-part article totaling nearly 150 pages.[67]

In 1934 Nice met both Konrad Lorenz and G. K. Noble—Lorenz in July at the Eighth International Ornithological Congress, held in Oxford, and Noble in December, at the Wilson Ornithological Club meeting in Pittsburgh. Although she was impressed by the paper Lorenz presented at the Oxford meeting and charmed by the bird stories he told when she and he had lunch together at the invitation of Julian Huxley, it does not appear that she identified Lorenz at the time as a man providing a new foundation for the understanding of bird behavior.[68] At any rate, in January 1935 Nice wrote to Noble saying that she was looking for a theoretical framework to provide structure for her researches. Animal psychologists in general, she indicated, did not impress her, but she found the views of Edward C. Tolman promising:

> the trouble with most animal psychologists is that they know only what their white rats do in the laboratory, and they never have studied intensively a wild animal in his natural environment.
>
> Tolman's "Purposive Behavior in Animals and Men" has been the most helpful material that I have found; it gives fundamentals, and I believe, the means for working out the total behavior. I am enclosing three attempts to work out an interpretation of Song Sparrow behavior on Tolman's lines: an outline of the statement in general; a table showing the strength of the different drives throughout their year, and also to some extent their mode of expression; and finally charts indicating the relative role of most of the drives at different significant times in the life of the birds.[69]

The striking effect of Lorenz's Kumpan article of 1935 on Nice's thinking is evidenced by the way Lorenzian ideas replaced Tolman's in her attempts to make sense of her extensive knowledge of the facts of bird behavior. She reviewed Lorenz's article in *Bird Banding* as soon as the article came out, lavishing praise upon it. Impressed by Lorenz's great fund of knowledge, his insistence on the diversity of behavior of different bird species, and his concept of releasers, she described the article variously as providing a "revolutionary and illuminating viewpoint," as "a most remarkable paper of fundamental importance," as a "great paper," and as a "solid foundation on which to build."[70]

Having corresponded with Lorenz from 1935 if not earlier, Nice arranged to follow the International Ornithological Congress in France in May 1938 with a month's visit to Lorenz's home in Austria, where she hoped to learn how to raise baby altricial birds and what to look for in their development. One result of her stay there was that she and Lorenz planned to write a joint article on "the maturation of some activities in young redstarts and serins," she providing the observations and he providing the theoretical discussion.[71] Later in the fall of the same year, at a special AOU symposium on "The Individual and Species" (together with Francis H. Herrick, G. K. Noble, F. C. Lincoln, and Niko Tinbergen), she willingly took as her subject "The Social Kumpan and the Song Sparrow." She began her paper with the observation that Lorenz's Kumpan article provided "a sure foundation for the study of bird behavior."[72]

It must not be supposed, however, that Nice was simply the passive recipient of guidance provided by others. She in turn posed instructive challenges to Lorenz, Noble, and other students of animal behavior by asking them pointed questions and making them confront each other's views and the behavioral evidence she knew from her own researches and her unrivaled knowledge of the ornithological literature. It was she who arranged the interaction of her old friend Wallace Craig and Lorenz, which proved of considerable importance for the evolution of Lorenz's thinking.[73] And it was she who set up the correspondence between Francis H. Herrick and Lorenz (Herrick then being the one who encouraged Lorenz to prepare an English summary of the Kumpan article for the *Auk*—an act on Herrick's part described in *Bird-Banding* as "the ornithological Good Deed of the decade").[74] She, along with Ernst Mayr and others, helped arrange Tinbergen's trip to the United States in 1938, a trip during which Tinbergen became personally acquainted with Noble, Yerkes, the young Daniel Lehrman, and other American investigators. It was also Nice who continually pressed G. K. Noble regarding his opinion of Lorenz's and Tinbergen's views.

However impressed Nice may have been by Noble telling her, when they first met, that he had a card catalog of 10,000 references on animal behavior, and however disproportionate were the resources he had at his disposal compared with those she had at hers—he with his major research facility and she with her badge identifying her as "Special Game Protector of the State of Ohio," to discourage the young boys of Columbus from shooting song sparrows—Nice was not intimidated by the large, hard-driving chairman of the Department of Experimental Biology at the American Museum.[75] To the contrary (at least in the late 1930s), she kept after Noble, making him clarify the points on which he found Lorenz's work lacking.

After Nice and Noble presented papers at the AOU symposium on "The Individual and the Species," they exchanged their papers in order to receive each other's comments. Noble informed Nice that while he regarded her paper as "splendid," he was unable to go along with her statement that Lorenz's theory of the Kumpan provided "a *sure* foundation for the study of bird behavior." Nice acknowledged in return that she had had to do her paper "in something of a rush" and that she had not been entirely comfortable with that choice of words herself. She proceeded to drop the word "sure" in the published version of her paper. However, in response to other comments from Noble, she expressed her agreement with Lorenz "that territory to a large extent nullifies dominance" and her skepticism concerning Noble's reluctance to believe that birds could be born with instinctive visual images of things. She also told Noble that certain experiments he had conducted did not in her view prove what he claimed for them, and further that his AOU paper had tried to cover too much ground in too short a time. In a later letter she underlined a question for Noble: "*Do you reject Lorenz' basic theory that striking colors and structures are releasers of instinctive behavior in other members of the species?*"[76]

Nice's efforts to identify the differences between the Noble and Lorenz camps and to get Noble and Lorenz to confront these differences were important ones. How profitable an interaction there might ultimately have been between Noble and Lorenz had Noble not died is difficult to say. Much more experimentally oriented than Lorenz—much more inclined, for example, to conceive of a means of *testing* whether imprinting differed from learning in being irreversible—Noble had legitimate questions concerning Lorenz's concepts of "imprinting," "inborn patterns," "releasers," and the like.[77] Lorenz in contrast had a more intimate knowledge of avian behavior under natural conditions and was a bolder theoretician. A sustained interaction between them would have been very significant for the development of animal behavior studies, had their respective personalities allowed it and had Noble's death not intervened.

While Nice provided suggestions, criticisms, and encouragement to other investigators through her extensive personal correspondence, she also made certain there was a public means of informing ornithologists of the latest work on bird behavior, ecology, life histories, and so forth. In 1934 she began to review the current ornithological literature (including that published in Europe) for *Bird-Banding*. Over the next eight years she wrote literally hundreds of abstracts of the latest work in the field. The way she signaled the importance of Lorenz's Kumpan paper has already been mentioned. Her enthusiasm for Lorenz's work, however, should not be taken as an indication that she was uncritical in her reviews. For example, she described F. B. Kirkman's book, *Bird Behaviour*, as "an important book containing a vast amount of material on Gull psychology," but she added, "the author fails to weave the different parts of the pattern into a comprehensive whole." Likewise she identified Allee's *The Social Life of Animals* as "an interesting and well-written book" but noted further: "The ornithologist will find much of value on fundamental viewpoints and on experiments in the laboratory, but little on the behavior of wild birds." On the other hand, she hailed a popular article by Tinbergen as "very fine exposition of bird psychology in simple, clear style with plenty of examples." Tinbergen's "On the Analysis of Social Organization among Vertebrates, with Special Reference to Birds" she identified as an article that "should be most carefully studied by all those interested in bird behavior." As for Tinbergen's and D. J. Kuenen's study of "the releasing and orienting stimulus situations of gaping movements in young thrushes," she called this "a brilliant piece of pioneer work." As American readers benefited from Nice's thorough combing of the international ornithological literature, they could not have failed to note her insistence on the importance of the work being done by the continental ethologists. An editorial in *The Wilson Bulletin* identified Nice's *Bird-Banding* reviews as "one of the most valuable contributions to ornithology now being regularly published."[78]

The most important of Nice's own articles were her 1933 historical review of the theory of territorialism, her series of articles in the 1930s on the song sparrow, and her two major monographs on the life history of the song sparrow—the first on population and the second on behavior—published in the *Transactions of the Linnaean Society of New York* in 1937 and 1943, respectively. Through these publications and her other efforts she established herself as one of the most important American ornithologists of the time, establishing in her work a model for future field studies. With regard to questions of behavior, she ended up endorsing a viewpoint that was, in her words, "fundamentally that of Lorenz, Heinroth, and Tinbergen, all of whom are more conversant with and more sympathetic with wild birds and other animals than are the laboratory psychologists." Her own contributions were not so much in the area of theory as the area of practice, with the latter involving not only her own remarkable field studies but also the various roles she played in the network of communication among twentieth-century investigators of bird behavior.

Of Nice's later writings, one of the most interesting, however brief, was her contribution on "The Question of Sexual Dominance" for a *Festschrift* for Erwin Stresemann. In this short article she shrewdly observed that Schjelderup-Ebbe, the Norwegian biologist who introduced the idea of social dominance to behavior studies, "was a man with a mission; he interpreted all life in terms of dominance, and particularly male dominance which he considered essential for the welfare of the world. . . . With such a bias it is no wonder that he hopelessly confused social and sexual dominance." Nice described how she herself had followed contemporary ideas of male dominance in her writings on song sparrows in the 1930s, but had begun to suspect that the male's behavior was not an expression of dominance but instead a signal of readiness to mate. Citing a posthumous paper by Oscar Heinroth denying that the male pigeon dominates his mate, Nice concluded that although social dominance could be observed and measured, the theory of sexual dominance had lost its chief support.[79] It is perhaps fitting that this observation came from the first woman to be made president of a major American ornithological society (the Wilson Ornithological Club, 1938–1939), and who had quite appropriately objected to being identified as a housewife rather than a trained zoologist.[80] Having been closed out of a professional scientific career by the gender demands of contemporary American society, Nice in a 1952 article on Althea Sherman expressed her sentiments on the situation: "Our highly-educated, gifted women have to be cooks, cleaning women, nurse maids. Men who could do notable research have their time wasted in mere routine. We who cherish things of the mind should face this evil and strive earnestly to give such men and women a chance to make the highest contribution to society of which they are capable."[81]

Conclusion

Between the two world wars, animal behavior studies in the United States were pursued on a variety of fronts. Allee's ecological and laboratory approach to behavior, Noble's combination of field, phylogenetic, and physiological studies, and Nice's field and home investigations of behavioral ecology and behavioral development all constituted important research initiatives which continued to be represented in American work after the end of the Second World War.

What American zoologists (and psychologists too, for that matter) still lacked after the war were the additional institutional structures needed to give animal behavior studies a clear identity. They needed to be able to have settings in which they could identify themselves first and foremost as students of animal behavior—rather than as ornithologists, herpetologists, entomologists, or indeed as zoologists or psychologists. Wondering "about the future of animal study as a discipline in psychology," T.C. Schneirla wrote to Robert Yerkes in 1944 about the possibility of reviving the *Journal of Animal Behavior*, noting the important contributions that could be made by a journal that would "assist a rapprochment [sic] of psychologists and zoologists interested in the more naturalistic aspects of animal study."[82] Yerkes's response, however, was discouraging, and major efforts to establish animal behavior journals after the war came from Europe and England, where Tinbergen (who told Yerkes that the Europeans lamented how *Journal of Animal Behavior*'s naturalistic style had been abandoned by *Journal of Comparative Psychology*) went to work immediately to replace the discontinued *Zeitschrift für Tierpsychologie* with a more international journal, *Behaviour,* and the *Bulletin* of the Institute for the Study of Animal Behavior became (in 1953) the *British Journal of Animal Behaviour.*[83]

As indicated above, a Committee for the Study of Animal Societies under Natural Conditions was formed in the United States in 1947. But it was not until 1963 that American zoologists concerned with behavior had a central umbrella organization of their own, the Animal Behavior Society. After the war the American Museum of Natural History did have its Department of Animal Behavior—however changed it may have been under new leadership. There and elsewhere in the United States behavioral researches continued, influenced by the work of Allee, Noble, and Nice, shaped by local circumstances, and also responding, in part sympathetically and in part critically, to the work of the continental ethologists. As for the continental ethologists—including the British, who had attracted Tinbergen to a position at Oxford—they had been more successful in their efforts to establish ethology as a recognizable scientific discipline. Interestingly enough, however, their efforts, too, were shaped by local circumstances—and also at least to some extent by challenges posed to them by their American counterparts.

Acknowledgments

The authors are grateful to the University of Chicago, Cornell University, the Rockefeller Archive Center, Yale University, and Charles Myers of the Department of Herpetology at the American Museum of Natural History for permission to use and quote from archival materials in their collections. Both authors are pleased to acknowledge research support from the National Science Foundation (Mitman for grant number SES-8603920 and Burkhardt for grant number SOC78-05922). Burkhardt also gladly acknowledges support from the Research Board of the University of Illinois at Urbana-Champaign, and Mitman expresses his thanks to the Rockefeller Foundation for support of this research.

Notes

1. Julian S. Huxley, "Bird-watching and Biological Science," *Auk,* 1916, *33:* 142–161, 256–270.

2. Richard W. Burkhardt, Jr., "The *Journal of Animal Behavior* and the Early

History of Animal Behavior Studies in America," *Journal of Comparative Psychology,* 1987, *101:* 223–230.

3. Philip J. Pauly, *Controlling Life: Jacques Loeb and the Engineering Ideal in Biology* (New York: Oxford University Press, 1987), pp. 80–81; Richard W. Burkhardt, Jr., "Charles Otis Whitman, Wallace Craig, and the Biological Study of Animal Behavior in the United States," in Ronald Rainger, Keith R. Benson, and Jane Maienschein, eds., *The American Development of Biology* (Philadelphia: University of Pennsylvania Press, 1988), pp. 185–218; and Philip J. Pauly, "The Loeb-Jennings Debate and the Science of Animal Behavior," *Journal of the History of the Behavioral Sciences,* 1981, *17:* 504–515.

4. V. E. Shelford, "Preliminary Note on the Distribution of the Tiger Beetles (Cicindela) and Its Relation to Plant Succession," *Biological Bulletin,* 1908, *14:* 9–14; idem, "Life Histories and Larval Habits of the Tiger Beetles (*Cicindelidae*). *Journal of the Linnean Society of London,* 1908, *30:* 157–184.

5. There was a strong tradition of plant ecology research at Chicago that had a marked influence on Shelford's dissertation research, but this influence is beyond the scope of this essay. For a discussion of the Chicago school of plant ecology, see Robert P. McIntosh, The *Background of Ecology: Concept and Theory* (Cambridge: Cambridge University Press, 1985), pp. 39–49; Ronald C. Tobey, *Saving the Prairies: The Life Cycle of the Founding School of American Plant Ecology, 1895–1955* (Berkeley: University of California Press, 1981), pp. 106–109; Donald Worster, *Nature's Economy: The Roots of Ecology* (Garden City, N.Y.: Anchor Press, 1979), pp. 206–208. The importance of physiology in the emergence of plant ecology is discussed in Eugene Cittadino, "Ecology and the Professionalization of Botany in America, 1890–1905," *Studies in the History of Biology,* 1980, *4:* 171–198; and Joel B. Hagen, "Organism and Environment: Frederic Clements's Vision of a Unified Physiological Ecology," in Rainger, Benson, and Maienschein, *The American Development of Biology,* pp. 257–280.

6. V. E. Shelford, "Physiological Animal Geography," *Journal of Morphology,* 1911, *22:* 551–618, on p. 594.

7. W. C. Allee, "An Experimental Analysis of the Relation Between Physiological States and Rheotaxis in Isopoda," *Journal of Experimental Zoology,* 1912, *13:* 270–344.

8. V. E. Shelford and W. C. Allee, "The Reaction of Fishes to Gradients of Dissolved Atmospheric Gases," *Journal of Experimental Zoology,* 1913, *14:* 207–263. For similar studies by other students, see Morris M. Wells, "The Reactions and Resistance of Fishes in Their Natural Environment to Salts," *Journal of Experimental Zoology,* 1915, *19:* 243–283; and C. F. Phipps, "An Experimental Study of the Behavior of Amphipods with Respect to Light Intensity, Direction of Rays, and Metabolism," *Biological Bulletin,* 1915, *28:* 210–223.

9. For biographical information on Allee, see Karl P. Schmidt, "Warder Clyde Allee, 1885–1955," *National Academy of Sciences. Biographical Memoirs,* 1957, *30:* 3–40; Alfred E. Emerson and Thomas Park, "Warder Clyde Allee, Ecologist and Ethologist," *Science,* 1955, *121:* 686–687; and W. C. Allee, "About Warder Clyde Allee," in J. R. Newman, ed., *What Is Science?* (New York: Simon and Schuster, 1955), pp. 228–230.

10. Lillie to Allee, 12 November 1920, Frank Rattray Lillie Papers (hereafter Lillie Papers), Box 1, Folder 3, Special Collections, University of Chicago.

11. For other historical accounts of Allee's professional career, see Edwin M. Banks, "Warder Clyde Allee and the Chicago School of Animal Behavior," *Journal of the History of the Behavioral Sciences,* 1985, *21:* 345–353; Joseph A. Caron, "La théorie de la cooperation animale dans l'écologie de W. C. Allee: analyse du double registre d'un discours" (master's thesis, University of Montreal, 1977); Gregg Mitman, "From the Population to Society: The Cooperative Metaphors of W. C. Allee and A. E. Emerson," *J. Hist. Biol.,* 1988, *21:* 173–194; idem, "Evolution by Cooperation: Ecology, Ethics, and the Chicago School, 1910–1950" (Ph.D. dissertation, University of Wisconsin, 1988).

12. "Courses of Instruction," Zoology Department Records, Box 4, Folder 1, Special Collections, University of Chicago.

13. W. C. Allee, *Animal Aggregations: A Study in General Sociology* (Chicago: Chicago University Press, 1931), p. 3.

14. A summary of these experiments is available in W. C. Allee, "Animal Aggregations," *Quarterly Review of Biology,* 1927, *2:* 367–398; idem, *Animal Aggregations*; idem, "Recent Studies in Mass Physiology," *Biological Reviews,* 1934, *9:* 1–48; and idem, *The Social Life of Animals* (New York: Norton, 1938).

15. W. C. Allee and J. F. Schuett, "Studies in Animal Aggregations: The Relationship between Mass of Animals and Resistance to Colloidal Silver," *Biological Bulletin,* 1927, *53:* 301–317, on pp. 315–316.

16. W. C. Allee, "Animal Aggregations: A Request for Information," *Condor,* 1923, *23:* 129–131, on pp. 129–130.

17. W. C. Allee, "Co-operation among Animals," *Chicago Alumni Magazine,* 1928, *20:* 418–425, on p. 423.

18. For a more detailed account of Allee's experiences as a liberal pacifist and the political implications of his research, see Mitman, "Evolution by Cooperation."

19. The tradition of embryological research at Chicago and the department's skepticism of genetics are touched upon in Scott F. Gilbert, "Cellular Politics: Just, Goldschmidt, Waddington and the Attempt to Reconcile Embryology and Genetics," in Rainger, Benson, and Maienschein, *The American Development of Biology,* pp. 311–346; Jane Maienschein, "Whitman at Chicago: Establishing a Chicago Style of Biology?" in Rainger, Benson, and Maienschein, *The American Development of Biology,* pp. 151–182; William B. Provine, *Sewall Wright and Evolutionary Biology* (Chicago: University of Chicago Press, 1986), pp. 168–169; and Jan Sapp, *Beyond the Gene: Cytoplasmic Inheritance and the Struggle for Authority in Genetics* (Oxford: Oxford University Press, 1987), pp. 9–12.

20. Lillie memo, 13 March 1934, Warder Clyde Allee Papers (hereafter WCAP), Box 6, Frank R. Lillie Folder, Special Collections, University of Chicago.

21. For a discussion of Lillie's research on the physiology of sex, see Clarke, this volume. For a history of the NRC Committee for Research in Problems of Sex which includes a list of grants received by Chicago, see Sophie D. Aberle and George W. Corner, *Twenty-Five Years of Sex Research: History of the National Research Committee for Research in Problems of Sex, 1933–1947* (Philadelphia: Saunders, 1953). Information on Lillie's grant is also available in RG 1.1, Series

200, Box 40, Folder 453, Rockefeller Archive Center, Tarryton, New York (hereafter RAC).

22. This change in funding is discussed in RBF memo, 16 February 1937, RG1.1, 216D, Box 8, Folder 108, RAC; Weaver to Lillie, 6 October 1933; Weaver memo, 8–11 September 1933; and Weaver memo, 6 November 1933, RG.1.1, 216D, Box 8, Folder 105, RAC. For Allee's initial reaction, see Allee to Lillie, 10 December 1934, WCAP, Box 6, Frank R. Lillie Folder.

23. See Yerkes to Allee, 23 March 1935; Allee to Yerkes, 11 March 1935; Memo Concerning NRC Grant for 1935–1936; Yerkes to Allee, 6 March 1936, WCAP, Box 15, Research Applications Folder. See also Yerkes to Weaver, 11 February 1936; Allee to Yerkes, 8 February 1936; Weaver to Yerkes, 26 February 1936, RG1.1, 200, Box 39, Folder 440, RAC.

24. Many of these experiments are summarized in W. C. Allee, "Social Dominance and Subordination among Vertebrates," in Robert Redfield, ed., *Levels of Integration in Biological and Social Systems* (Lancaster: Jacques Cattell Press, 1942), pp. 139–162; and idem, "Dominance and Hierarchy in Societies of Vertebrates," in *Structures et physiologie des sociétés animales,* Colloques Internationaux du Centre National de la Recherche Scientifique, 1950–1952, *34:* 151–181.

25. Allee to Lillie, 5 August 1938, WCAP, Box 6, Frank R. Lillie Folder. Allee's seminal article on the subject of group selection is W. C. Allee, "Concerning the Origins of Sociality in Animals," *Scientia,* 1940, *67:* 154–160. For a discussion of population selection as a fundamental aspect of Chicago ecology, see Mitman, "From the Population to Society," and William Kimler, "Advantage Adaptiveness, and Evolutionary Ecology," *J. Hist. Biol.,* 1986, *19:* 215–233.

26. Nikolaas Tinbergen, *Social Behavior in Animals with Special Reference to Vertebrates* (New York: Wiley, 1953), p. 71.

27. These included John T. Emlen and F. W. Lorenz, "Pairing Responses of Free-Living Quail to Sex Hormone Implants," *Auk,* 1942, *59:* 369–378; Walter E. Howard and John T. Emlen, Jr., "Intercovey Social Relationships in the Valley Quail," *Wilson Bulletin,* 1942, *54:* 162–170; E. P. Odum, "Annual Cycle of the Black-Capped Chickadee. I," *Auk,* 1941, *58:* 314–333; and J. W. Scott, "Mating Behavior of the Sage Grouse," *Auk,* 1942, *59:* 477–498.

28. For a review of these studies, see Margaret M. Nice, "The Role of Territory in Bird Life," *American Midland Naturalist,* 1941, *26:* 441–487.

29. These included J. P. Scott, A. M. Guhl, N. E. Collias, and E. B. Hale.

30. Biographical information on Noble is available in Clifford H. Pope, "Gladwyn Kingsley Noble," *National Cyclopaedia of American Biography,* 1944, *31:* 396; William K. Gregory, *Copeia* (1940), 274–275; "Gladwyn Kingsley Noble," Gladwyn Kingsley Noble Papers (hereafter GKNP); "Noble, G. K.: Biography II" Folder, Department of Herpetology, American Museum of Natural History.

31. Noble's importance and role in herpetology systematics are discussed by Ogilvie, this volume.

32. On Gregory, see Ronald Rainger, "Vertebrate Paleontology as Biology: Henry Fairfield Osborn and the American Museum of Natural History," in Rainger, Benson, and Maienschein, *The American Development of Biology,* pp. 219–256.

33. Osborn to Noble, 17 March 1928, GKNP, "AMNH Presidency: Henry F. Osborn I" Folder. See also Noble to Osborn, 20 March 1928, GKNP, "AMNH

Presidency: Henry F. Osborn I" Folder and Sherwood to Noble, 18 May 1928, GKNP, "Department History" Folder. On Morgan's move to Caltech, see Garland E. Allen, *Thomas Hunt Morgan: The Man and His Science* (Princeton: Princeton University Press, 1978), pp. 334–347.

34. Noble to Sherwood, 22 May 1928, GKNP, "Noble, G. K.: Biography I" Folder; Sherwood to Noble, 1 June 1928, GKNP, "Department: Budgets" Folder.

35. "Laboratory of Experimental Biology" manuscript, 19 May 1933, GKNP, "AMNH Departments: Experimental Biology" Folder.

36. G. K. Noble and L. B. Richards, "The Induction of Egg-laying in the Salamander, Eurycea bislineata, by Pituitary Transplants," *American Museum Novitates,* 1930, n. 396.

37. For a brief description of the laboratory, see Noble to Burden, 15 October 1934, GKNP, "AMNH Departments: Experimental Biology" Folder. On the political and social context of the African Hall and the American Museum in general, see Donna Haraway, "Teddy Bear Patriarchy: Taxidermy in the Garden of Eden, New York City, 1908–1936," *Social Text,* 1984/85, *11:* 20–64.

38. See Sherwood to Noble, 13 January 1933; "Draft of Budget for 1934," 13 November 1933, GKNP, "Department: Budgets" Folder; and Faunce to Noble, 9 April 1935, GKNP, "AMNH Departments: Experimental Biology" Folder.

39. An account of WPA help in the department and the duties that Noble assigned to these individuals can be found in GKNP, "Department: Personnel (WPA)" Folder.

40. Noble to Burden, 15 October 1934, GKNP, "Departments: Experimental Biology" Folder.

41. Noble to Burden, 18 June 1931, GKNP, "Burden, W. Douglas" Folder.

42. Burden to Noble, n. d., GKNP, "Burden, W. Douglas" Folder.

43. G. K. Noble, "Recent Advances in Our Knowledge of Sex," *New York World-Telegram,* June 1931.

44. G. K. Noble, *The Biology of the Amphibia,* 2nd ed. (Dover, 1954), pp. 377 and 390.

45. Noble's laboratory received a total of $23,500 from the Josiah Macy, Jr. Foundation and $11,900 from the National Research Council Committee for Research in Problems of Sex between the years 1935 and 1940. See GKNP, "Foundations & Institutes: National Research Council" Folder; "Foundations & Institutes: Josiah Macy, Jr." Folder; and budget expenditure statements in "AMNH Departments: Experimental Biology" Folder.

46. G. K. Noble and Brian Curtis, "The Social Behavior of the Jewel Fish, *Hemichromis Bimaculatus* Gill," *Bulletin of the American Museum of Natural History,* 1939, *76:* 1–46.

47. G. K. Noble, "The Experimental Animal from the Naturalist's Point of View," *American Naturalist,* 1939, *73:* 113–126, on p. 122. The overall scope of Noble's research is evident from progress reports to the foundations. See GKNP, "Foundation & Institutes: National Research Council" Folder, and "Foundations & Institutes: Josiah Macy, Jr." Folder.

48. G. K. Noble, M. Wurm, and A. Schmidt, "Social Behavior of the Black-Crowned Hight Heron," *Auk,* 1938, *55:* 7–40.

49. Part of the hormonal research was published in G. K. Noble and M. Wurm,

"The Effect of Testosterone Proprionate on the Black-Crowned Night Heron," *Endocrinology,* 1940, *26:* 837–850.

50. Noble to Yerkes, 31 August 1939, GKNP, "Foundations & Institutes: National Research Council" Folder.

51. On the hiring of Beach, see Noble to Beach, 1 November 1937, GKNP, "AMNH Departments: Experimental Biology" Folder.

52. Noble, "The Experimental Animal," pp. 123–124.

53. Noble to Fremont-Smith, 15 February 1937, GKNP, "Foundations & Institutes: Josiah Macy, Jr." Folder.

54. Frank Fremont-Smith of the Macy Foundation had Noble's translation of Lorenz's paper mimeographed and sent out to a number of psychologists. Noble himself also sent the paper out to biologists working on animal behavior. See GKNP, Fremont-Smith to Noble, 6 May 1937, "Foundations & Institutes: Josiah Macy, Jr." Folder; Noble to Davis, 1 October 1937, "Davis, David E.: Biological Labs, Harvard" Folder; Allee to Noble, 2 January 1940, "Universities & Colleges: US-Chicago" Folder.

55. Konrad Lorenz, "Der Kumpan in der Umwelt des Vogels," *Journal für Ornithologie,* 1935, *83:* 137–213, 289–413.

56. For the offer to Tinbergen of the position at the Huyck Preserve, see Noble to Tinbergen, 20 February 1939 and 12 December 1939, GKNP, "Tinbergen, N." Folder.

57. The identification of Noble's affliction as "Ludwig's quinsy" is from Walter L. Necker, "Gladwyn Kingsley Noble, 1894–1940: a Herpetological Bibliography," *Herpetologica,* 1940, *2:* 47. On Chapman's assumption that the department would need to be dissolved after Noble's death, see Frank A Beach, "Frank A Beach," *A History of Psychology in Autobiography.* 1974, *6:* 50. The American Museum school's critique of European ethology found its historically most important expression in Daniel Lehrman, "A Critique of Konrad Lorenz's Theory of Instinctive Behavior," *Quarterly Review of Biology,* 1953, *28:* 337–363.

58. Birds have proved to be especially important subjects for behavior study in the twentieth century. The concepts of territory, peck-order, and imprinting were all derived from ornithological studies. See Ernst Mayr, "The Role of Ornithological Research in Biology," in *Proceedings of the XIIIth International Ornithological Congress, Ithaca, 17–24 June 1962* (Baton Rouge, 1963), pp. 27–38. In the United States in the second decade of the twentieth century, birds represented the class in which there was the greatest overlap between psychologists and biologists in terms of organism choice (see Burkhardt, "The *Journal of Animal Behavior* and the Early History of Animal Behavior Studies in America").

59. See Richard W. Burkhardt, Jr., "Julian Huxley and the Rise of Ethology" (in press).

60. For biographical information on Margaret Morse Nice, see especially Margaret Morse Nice, *Research Is a Passion with Me* (Toronto: Consolidated Amethyst Communications, 1979) and Milton B. Trautman, "In Memoriam: Margaret Morse Nice," *Auk,* 1977, *94:* 430–441. The importance of gender-dictated expectations for the course of Margaret Morse Nice's career is discussed by Laurel Furumoto and Elizabeth Scarborough, "Placing Women in the History of Comparative Psychology: Margaret Floy Washburn and Margaret Morse Nice," in Ethel Tobach, ed., *Histori-*

cal Perspectives and the International Status of Comparative Psychology (Hillsdale, N.J.: Lawrence Erlbaum, 1987), pp. 103–117, and by Marianne Gosztonyi Ainley, "Field Work and Family: North American Women Ornithologists, 1900–1950," in Pnina G. Abir-Am and Dorinda Outram, eds., *Uneasy Careers and Intimate Lives* (New Brunswick: Rutgers University Press, 1987), pp. 60–76.

61. Nice, *Research Is a Passion with Me,* p. 33.

62. On American ornithology, including a discussion of the role of ornithological societies, see Ernst Mayr, "Epilogue: Materials for a History of American Ornithology," in Erwin Stresemann, *Ornithology from Aristotle to the Present* (Cambridge, Mass.: Harvard University Press, 1975), pp. 365–396.

63. Margaret Morse Nice, "Some Letters of Althea Sherman," *Iowa Bird Life,* 1952, *22:* 51–55.

64. Nice, *Research Is a Passion with Me,* p. 90.

65. Ibid., p. 94.

66. For example, after the Wilson Club meeting at the University of Michigan in 1928, she noted: "Friends are the best part of a bird meeting. The Museum was an inspiration. It was pleasant to get acquainted with Dr. Josselyn Van Tyne, Curator of Birds at the University Museum." Nice, *Research Is a Passion with Me,* p. 93.

67. Margaret Morse Nice, "Zur Naturgeschichte des Singammers," *Journal für Ornithologie,* 1933, *81:* 552–595; 1934, *82:* 1–96.

68. Margaret Morse Nice, "My Debt to Konrad Lorenz," *Zeitschrift für Tierpsychologie,* 1963, *20:* 461.

69. Nice to Noble, 25 January 1935, GKNP, "N—Miscellaneous Correspondence" Folder.

70. Margaret Morse Nice, "The *Kumpan* in the Bird's World. The Fellow-member of the Species as Releasing Factor of Social Behavior," review in *Bird-Banding,* 1935, *6:* 113–114, 146–147.

71. Margaret Morse Nice, "Studies in the Life History of the Song Sparrow II. The Behavior of the Song Sparrow and Other Passerines," *Transactions of the Linnaean Society of New York,* 1943, *6:* 3.

72. Margaret Morse Nice, "The Social Kumpan and the Song Sparrow," *Auk,* 1939, *56:* 255–262. As noted below, Nice omitted the word "sure" in the published version of her paper.

73. Craig sent a set of his reprints to Lorenz, and Lorenz replied with a seven-page letter, according to Nice, *Research Is a Passion with Me,* pp. 140–141.

74. Ibid., p. 141. Thomas T. McCabe, "The Companion in the Bird's World," *Bird-Banding,* 1937, *8:* 183.

75. Margaret Morse Nice, "Studies in the Life History of the Song Sparrow. I," *Transactions of the Linnaean Society of New York,* 1937, *4:* 16. Nice expressed dissatisfaction with her lack of support in Ibid., p. 209, and p. 215: "A very interesting study of the population of Interpont could have been made if there had been a number of students and institutional support."

76. Noble to Nice, 24 October 1938; Nice to Noble, 28 October 1938 and 13 November 1938, Margaret Morse Nice Papers, (hereafter MMNP), Library, Cornell University. This kind of questioning continued in later letters. See also Nice to Lorenz, 5 October 1938; Lorenz to Nice, 9 November 1938 and 14 February 1939; and Nice to A. L. Rand, 4 December 1940, MMNP.

77. See Noble to Lorenz, 26 July 1936, GKNP, "Lorenz, Konrad" Folder. Also Noble to Nice, 24 October 1938, 31 October 1938, 18 November 1938, 14 November 1940, and Nice to Lorenz, 29 January 1939, MMNP.

78. For the reviews cited here, see *Bird-Banding,* 1937, *8:* 187; 1938, *9:* 218; 1939, *10:* 99, 136, 174. See also "Editorial," *The Wilson Bulletin,* 1936, *47:* 144.

79. Nice, "The Question of Sexual Dominance," in Ernst Mayr and Ernst Schuz, eds., *Ornithologie als biologische Wissenschaft. Festschrift zum 60. Geburtstage von Erwin Stresemann* (Heidelberg: C. Winter, 1949), pp. 158–161.

80. See Trautman, "In Memoriam: Margaret Morse Nice."

81. Margaret Morse Nice, "Some Letters of Althea Sherman," *Iowa Bird Life,* 1952, *22:* 55.

82. T. C. Schneirla to R. M. Yerkes, 8 March 1944, Robert M. Yerkes Papers (hereafter RMYP), Library, Yale University.

83. Tinbergen to Yerkes, 18 October 1946, RMYP. See also John Durant, "The Making of Ethology: The Association for the Study of Animal Behaviour, 1936–1986," *Animal Behaviour,* 1986, *34:* 1601–1616.

Sharon E. Kingsland

8

Toward a Natural History of the Human Psyche: Charles Manning Child, Charles Judson Herrick, and the Dynamic View of the Individual at the University of Chicago

The molecular revolution that followed the discovery of DNA's double helical structure has colored our understanding of American biology between the wars. Understandably, historians have searched at the boundary between biology and the physical sciences for the roots of this revolution, tracing it to the emergence of new fields of research that turned much of biology into a science of molecules rather than of organisms. Whereas we know a great deal about biology in relation to the physical sciences, we know much less about biology's relation to the social sciences between the wars. Yet at this boundary also there was active discipline building, a sign of growing interest in human biology outside the context of medical science. At this boundary biologists competed with social scientists for authority to address pressing social problems.

Evolutionary biologists had long believed that the biological method, based on comparative study of humans and lower animals, could provide an objective basis for the discussion of human society. At the start of the twentieth century, the goal of solving social problems through biology had been the justification for the eugenics movement, which culminated in the emergence of human genetics during the 1930s.[1] In another essay in this volume, Diane Paul analyzes one such research program in the context of the Rockefeller Foundation's patronage of science. In addition to increased interest in human genetics between the wars, there were several related attempts to encourage a *biological* approach to the human species, or, in other words, a comparative approach to human biology that was distinct from both medicine and the social sciences.[2]

Stephen J. Cross and William R. Albury have drawn our attention to the way biologists participated in debates about the stability of American society between the wars.[3] They point out that one of the responses to technological and social change after the First World War was to look for natural conditions of stability in society, to discover ways to foster what were considered to be natural, organic processes of human collaboration. They examine the different ways that Hippocratic models of physiological regulation were applied by Walter B. Cannon and Lawrence J. Henderson to problems of social stability, each application reflecting the different political views of these biologists. Cannon merely extended biological theory to the social realm without intending to develop a new field; Henderson was more interested in discipline building and tried to create a synthetic "human biology," merging biology and the social sciences.

Other examples of overlap between biology and the social sciences abound. At the Johns Hopkins School of Hygiene and Public Health, Raymond Pearl developed a comparative demographic approach to the study of groups, both human and animal.[4] His research on the biological causes of population growth and decline led him into heated controversies with social scientists. In this volume Garland E. Allen writes about the connection between Pearl's population interests and the eugenics movement. The growth of primate behavior studies, spearheaded by Robert Yerkes at Yale University, represents another such attempt to tackle problems of human behavior from a biological perspective.[5] The strength of behaviorist psychology in the 1920s rested largely on its claims to ground psychology in biology and to develop an "objective" method of studying behavior that eliminated reference to introspective experience.[6] At the University of Chicago ecologists and evolutionary biologists analyzed the biological basis of social behavior through comparative study of lower organisms.[7] Some attempts to develop a new human biology were abortive, a notable instance being the efforts of mathematician Alfred J. Lotka to define in the 1920s a field called "physical biology" that included the study of human consciousness and behavior in a new science organized around the analysis of energy transformations within the biosphere.[8]

To reveal the pluralistic nature of the new human biology during these years, we would need several parallel histories of individual research strategies. The present essay is only one of these stories. The larger story is of interest not only for what it might tell us about the institutional and social politics of science in the interwar years, but also for what it would reveal about the variety of American attitudes toward "man the animal," attitudes which helped to shape the fields of psychobiology and sociobiology that emerged after the Second World War.

In this essay I focus on an approach to human biology that drew its methods from physiology and ecology and contributed primarily to the growth of comparative psychobiology and secondarily to what we may conveniently call "sociobiology" (which differs from modern sociobiology in lacking population genetics). This research program was centered at the University of Chicago and was fostered by the interdisciplinary atmosphere of the university. The essay will explore the attempts, developed in collaboration by Charles Manning Child (1869–1954) and his colleague Charles Judson Herrick (1868–1960), to create a new dynamic biology and then to turn this dynamic approach toward the study of human behavior and society. I consider Child's and Herrick's stories to be part of the history of modern sociobiology, a part that has been neglected largely because theirs was a different kind of synthesis than the blend of population genetics, population ecology, and ethology that Edward O. Wilson defined in the 1970s.[9] If we think of sociobiology broadly not as the discipline that Wilson defined but as an evolving debate about human social behavior and its biological origins carried on in the various fields of biology that bordered on the social sciences, we can begin to construct a fuller history of sociobiology that will reveal the complex relationship between biology and American culture.

Although my case study will concentrate on these two biologists, I shall briefly discuss the parallel views of their contemporary Herbert Spencer Jennings (1868–1947), biologist at Johns Hopkins University. I do this chiefly to elucidate the liberal political stance that was expressed in the biology I discuss, because I believe that Jennings articulated especially clearly the liberal philosophy that was implicit in Herrick's writings. My further purpose is to show the biologist of the interwar years in a role that was controversial at the time but which describes an important feature of interwar biology: the role not of the narrow specialist but of a participant in a larger debate about the meaning of human life and the problems of a democratic society.

The idea that the biologist had something relevant to say about social or political problems was controversial. Some biologists worried that involvement in social issues would compromise the "objectivity" of their science. Any involvement in philosophical or social debate threatened to reverse the professionalization process that had been under way since the late nineteenth century. Second, social scientists opposed biologists' claims to be authoritative in human affairs. They believed that human social problems were not essentially biological but were environmental and cultural in origin. Their environmentalist perspective was in part a justified reaction to the biological determinism of the eugenic enthusiasts, but it was also in part an assertion of disciplinary autonomy, as Hamilton Cravens has argued.[10] But some biologists were interested in blurring these boundaries between biology and the social sciences. Biologists, that is, wanted to address themselves to human problems, not necessarily by studying human beings exclusively or even directly, but by comparative study of the lower animals. The work of Child, a physiologist in the Department of Zoology at Chicago, and Herrick, a neurobiologist in the Department of Anatomy, represents a spirit of rapprochement between biology and the social sciences.[11]

In assessing their motives, we need to consider the impact of the First World War on biologists' perceptions of their roles as social critics. Gregg Mitman has argued that the war focused biologists' attention on the need to address the problem of the biological basis of democracy in a sustained manner.[12] Many biologists, men such as William Patten, Edwin G. Conklin, David Starr Jordan, and William Emerson Ritter, were pushed by the war into new roles as "philosophers of democracy," in which they articulated views about the biological origins of sociality. I believe Mitman is right to emphasize the war's role in sensitizing biologists to their larger responsibilities as social critics. In this essay I refer briefly to the possible effects of the war on Child and Herrick, but with the idea that closer analysis might reveal a more profound impact than I have indicated.

I shall also argue that the origins of this spirit of rapprochement, at least as far as Child and Herrick were concerned, can be traced farther back in the early twentieth century, when Americans were making biology over in a new spirit of professionalism. For this reason, though the essay will stress the interwar period, it will be necessary to begin our account around the turn of the century. The origins of psycho- and sociobiology can only be understood in the context of the efforts to create a new, dynamic, experimental biology around the turn of the century, a biology formed from a genuine synthesis of morphology and physiology.

Therefore this essay is also a contribution to a continuing debate about the transition at the turn of the century from natural history to experimental biology, a transition that produced what Garland Allen called the "naturalist-experimentalist dichotomy."[13] His thesis has encouraged others to make a closer analysis of the professionalization process, which reveals, however, an absence of any clear-cut "dichotomy" within the community of biologists: the story is more complex than his characterization indicates.[14] This essay confirms the difficulties of dividing biologists too sharply into opposed camps of naturalists and experimentalists, though the move toward an experimentally rigorous science was certainly part of the process of professionalization at the turn of the century.

Child's and Herrick's approach to the biological basis of human nature originated in an attitude toward biology that was essentially that of the natural historian, refined by the experimental method that shaped the new biology emerging at the turn of the century. The natural history viewpoint was challenged by the reductionist physicochemical biology and behaviorist psychology that gained in importance in the 1920s, but it was never wholly eclipsed by this approach. The naturalist's perspective underlay the emphasis on "holism" which cropped up repeatedly in the literature of this period and which was a prominent theme of Child's and Herrick's writings.

Part of the purpose of this essay is to discuss the meaning and origins of this holistic approach in relation to competing methods of the period, and to show how Child's and Herrick's way of thinking about biology reflected the academic environment at Chicago. One of the points I wish to emphasize is that the meaning of "holism" ultimately was both scientific and political: in the cases discussed here holistic statements that emphasized the relations between individual and environment, and which made rhetorical use of such concepts as "emergent evolution," reflected a progressive liberal viewpoint (though holism is not always an expression of liberal political views).

Two definitions are in order. By "holism," I refer not to a unified doctrine but to a research strategy that focused on the organism in its complex relations with the environment and which resisted modern attempts to reduce biology to chemistry and physics. The best way to define a given instance of holistic thinking is through an example; its meaning in this context should be clearer by the end of the essay. The term itself is imposed by me: both Child and Herrick would have called their biology "mechanistic," which for them meant a rigorously experimental science opposed to vitalistic theories that placed biological processes beyond the reach of science. But they would also have recognized a distinction between their approach to science and the more reductionist research strategies of such physicochemical biologists as Jacques Loeb. Their label "mechanistic" is confusing because we often equate mechanistic and reductionist science; therefore I shall call them holists.[15] In Child's case, holistic biology also implied an opposition to static, morphological approaches, because Child associated morphology with the speculative theorizing of the nineteenth-century German evolutionists that, like vitalism, appeared to him to be beyond the reach of experimental test. In Herrick's case, holism implied an opposition to radical behaviorism, because the psychological technique of introspection (or self-analysis) was excluded from the behaviorist's research.[16]

Just as one cannot easily pinpoint a meaning of "holism," but must see it as a research strategy evolving in response to reductionist and vitalistic arguments, so the sense of "liberalism" which I associate with certain holistic statements is not a fixed doctrine. Liberalism as an expression of faith in individual freedom was also evolving in response to technological change and to the challenges of socialism and fascism. In this essay I focus on certain key ideas of liberalism and their biological expression. These ideas included the belief that human nature—the mind and the character—was not fixed but was shaped throughout life by interaction with the social environment; a faith in gradual evolutionary progress rather than revolutionary or violent change; and a belief that the creative intelligence, which reached its highest expression in the cooperative activity of science, could solve social problems. This essay explores the way biology was used to support a faith in individual freedom and gradual social progress, a progress to be achieved by cooperative action rather than competitive struggle. Despite its vagueness, I use the term "liberalism" in describing this creed, in order to stress the connection between biology and the larger political debate between the wars.[17]

The Unity of the Organism

Charles Manning Child grew up in Connecticut and received his bachelor and master's degrees in 1890 and 1892, respectively, from Wesleyan University at Middletown, Connecticut. At age twenty-three he went to the University of Leipzig, where he spent one semester in Wilhelm Wundt's laboratory of experimental psychology, before settling into doctoral research under Rudolph Leuckart, a morphologist. Child stopped at the Zoological Station at Naples before returning to the United States, where in 1895 Charles Otis Whitman offered him a position as zoological assistant at the newly founded University of Chicago. There he built a distinguished career in physiology and invertebrate zoology. He became full professor in 1916 and was at Chicago until 1937, when he retired to Stanford University as a guest of the zoology department there.[18]

Although Child's thesis had been on a purely morphological subject in entomology, his great passion was experimental zoology. His writing in the early twentieth century shows a clear reaction to what he perceived to be a static, morphological approach to the problem of form, one that left the basic problems in biology unsolved. In the 1890s and early twentieth century, it was the embryologist who held center stage in biology: studies of growth, regeneration, and development absorbed the attention of leaders in biology and generated many controversial interpretations of the problem of organic unity.[19] How did the organism achieve pattern, order, and correlation of its activities? Child saw that to understand this problem one had to go beyond descriptive morphology and see the individual in dynamic terms, constantly reacting with its surroundings. The enunciation of the dynamic conception of the individual, around the broad intellectual problem of organismic pattern, was the central theme of his writings. Child's ideas and attitudes were characteristic of American experimental biology at the turn of the century.[20] His dynamic viewpoint grew out of a series of experiments that began in the 1890s on regeneration and development in lower organisms. By 1911, his ideas on organismic unity were well developed; they were articulated in two books published in 1915, *Senescence and Rejuvenescence* and *Individuality in Organisms*. Both of these books, summarizing fifteen years of research, expressed his enthusiasm for an experimental style of biological research, one that firmly repudiated not only vitalistic theories but also what Child thought were half-baked mechanistic explanations that merely echoed the style of nineteenth-century speculative morphology.

The problem of individual unity had been tackled in different ways, which Child identified loosely as "mechanistic" or "vitalistic," in the nineteenth century.[21] The most recent expression of the vitalist viewpoint grew from the biological research of Hans Driesch, who from 1895 argued that the organism was subject to laws peculiar to living things.[22] He named the controlling factor of life "entelechy," signifying an unknown, nonmaterial organizing principle. Driesch's vitalism was not necessarily opposed to current materialist theories, such as the material theory of heredity; it merely added the power of entelechy onto any theory of the material basis of life. The main goal of this approach was to express faith in a spiritual principle underlying reality and to reclaim a metaphysical role for the biologist.[23]

Child agreed with Driesch that it was impossible to deny that some ordering or controlling principle existed in the organism, but he rejected the way vitalism placed this principle beyond the reach of science. Child considered science to be mechanistic, by which he meant governed by a rigorous experimental method. He was aware that there were a number of mechanistic research programs in the nineteenth century, programs that tried to analyze the organism along physicochemical lines, as though it were a machine having a preestablished harmony in its construction. The nagging question for the mechanist was whether the organism's harmonious design could be explained by mechanistic logic.

For Child, the problem of unity could not be solved by investigating the structure of the organism alone. Instead one had to see the individual in relation to its environment: pattern and unity originated in this basic dynamic relationship between individual and its immediate environment. This dynamic conception challenged the dominant morphological view of the organism, substituting in its place a physiological interpretation that constantly referred structure back to function. The individual (a term which could refer to the whole organism or to a part) was neither solely the structural system studied by the morphologist nor the system of chemical reactions studied by the physiologist, but was a complex of dynamic changes occurring in a protoplasmic substrate, which was itself altered by these changes.[24]

Child expressed his dynamic concept of the individual by drawing on a metaphor, which he introduced in 1911: the organism was comparable neither to a crystal nor to a machine, two common metaphors of the time, but was more like a flowing stream.[25] The flow of water was the dynamic process, comparable to the current of chemical energy flowing through the organism. The banks and bed were the morphological features. In the river, "structure and function are connected as in the organism: the configuration of the channel modifies the morphology of the channel by deposition at one point, giving rise to structures such as bars, islands, flats, and by erosion at another. . . . The most important point for present purposes is that in the river, as in the organism, structure and function are indissociable and react upon each other."[26] Morphology and physiology were inseparable.

How, then, did protoplasmic pattern itself originate? In this dynamic system, Child believed he had found the starting point of organization in his discovery that different parts of the individual had different metabolic rates.[27] He thought that the rate of metabolism followed a well-defined gradient from a zone of high metabolic rate at the apical or head region to a low metabolic rate in the tail region. This primary axial gradient (or gradient along the anterior-posterior axis) could serve as a focus of organization because regions of high metabolic rate also dominated regions of lower rate. In other words, the head region would dominate the lower regions of the body. In every organism, all the processes of growth, development, regeneration, and reproduction were controlled by these relations of dominance and subordination between areas having different metabolic rates. Only the head region was truly independent or self-determining; anything happening in the posterior regions was determined and guided by the head region.

The crucial point was that these gradients would have been created by the local action of *external* factors. In studying the individual cell or organ, these factors might be external only to the part in question, not to the organism as a whole. Ultimately, however, the origin of individuality in evolutionary history could be traced to the relations between living protoplasm and the external world. In some cases the gradients were transitory, appearing and disappearing at different times. In other cases the gradients might persist even after the environment changed; that is, they would become hereditary. The organism as a whole, conceived in these dynamic terms, was a complex of metabolic gradients along differently oriented axes: together they combined and overlapped, as Child said, like the waves from several pulses crisscrossing in a wave trough.[28]

What was actually transmitted from the dominant to the subordinant regions were not chemical hormones but a general "excitation." Although Child was not opposed to the hormone theory as an explanation of how different parts of the organism were correlated with one another, he did not think that the *origin* of this correlation could be explained by the movement of chemicals. The dominance relationship depended instead on a transmitted change, the rudimentary transmission of energy that would later be perfected in the nervous system. Therefore the nervous system was the expression, in an advanced, specialized structure and function, of the basic condition that was the starting point of individuation.[29] In other words, protoplasm had the capacity for transmission of excitation, so that the appearance of the nervous system did not involve the origin of a *new* function that was different from the basic activities of protoplasm. In 1921 Child developed this thesis that the nervous system was not something arising de novo in evolution but was a system arising from the basic properties of the protoplasm.[30] Against the currently popular hormone theory (the subject of much research by his colleagues in Chicago), Child reinstated the nervous system as the most important physiological system in the body.

Child's approach to the problem of pattern can best be appreciated in contrast to that of Jacques Loeb. Loeb's argument, which appeared in 1916, was that ultimately one could understand all processes by understanding the flow of chemicals within the body.[31] He was trying to eliminate all teleological reasoning from biology by showing that the organism reacted with the machinelike precision that one finds in physics and chemistry. In effect, he argued that the whole issue of the "organism as a whole" was a pseudo-problem: there was no "whole" apart from the sum of the parts. But in explaining why the organism exhibited regular development, Loeb could only suggest that the pattern was predetermined in the structure of the egg. From Child's point of view, he had not solved the problem of explaining the origin of pattern; he had merely evaded the issue.[32]

Loeb in turn found Child's ideas to be unnecessarily complicated. Although Child believed that he had discovered gradients in a wide variety of lower organisms, his theory had an ad hoc quality that he never fully resolved. The question always remained as to how real these gradients were and how effectively they could explain the dominant-subordinate relationship between parts.[33] But Child regarded his theory as at least a good working hypothesis, one that had the merit of locating the origin of pattern in the relations between protoplasm and environment. The benefit of the gradient theory was its quantitative aspect: the gradient slope and rate of metabolism could be measured exactly, and this precision, Child thought, gave it an advantage over qualitative concepts in physiology. The method he used to locate these gradients was to kill the organism slowly by placing it in a lethal solution such as cyanide and watching the process of disintegration and death. The order of disintegration of tissue revealed the gradient, he believed, because he assumed that the most active parts would be most susceptible to the toxic agent and the least active parts the least susceptible. Much of Child's later research was devoted to showing the existence of these gradients in various embryonic and adult organisms.[34]

The Ecological Concept of the Individual

Child's efforts to explain the origin of pattern were compatible with a Lamarckian theory of evolutionary progress.[35] As we shall see, Child also believed that processes at work in the biological organism were the same as those operating in society, thereby allowing one to discuss social evolution as part of the general process of biological evolution. In this respect Child's ideas found echoes within the social sciences at the University of Chicago. Philip J. Pauly has pointed out that liberal Protestant academics at the University of Chicago tended to accept some form of progressive evolutionism, whether leaning toward neo-Lamarckian or orthogenic mechanisms, and ideas about biological evolution were often extended in the 1890s to discussions about human progress within the newly emerging social sciences.[36] It is not clear how directly Child may have been influenced by social thought in this period, but his ideas were in the same vein. He was not a dogmatic Lamarckian, though he did accept the possibility of the inheritance of acquired characteristics as late as the 1910s and his "dynamic conception" started from the same recognition that the individual was in interaction with its surroundings.[37] By the 1920s, when Lamarckian ideas were strongly challenged in the United States, he distanced himself from this or any other theory of evolution, claiming that his work was mainly physiological and not addressed to historical questions.[38] Even his evolutionary discussions, he claimed, did not depend on an acceptance of Lamarckian hypotheses. He was, however, always thinking of evolution as a progressive process.

Child's dynamic conception of the individual also reflected his more immediate environment, in this case the ideas being developed by contemporary ecologists, including Child's colleagues at Chicago. Charles C. Adams, who taught a course in animal ecology at Chicago as early as 1902 and received his Ph.D. from Chicago in 1908, believed that Child's ideas were basically adaptations of ecological concepts.[39] Adams himself was enunciating in the 1910s a distinctive "dynamic-process" viewpoint in ecology, which borrowed ideas from physical chemistry as well as geology. At Chicago, Henry Chandler Cowles was the main advocate of a dynamic geological approach to plant ecology, and it is likely that contact with Cowles would have made Child aware of current ecological theory.[40] But Child also had a direct influence on ecological thought, especially through Victor Ernest Shelford, who studied the Indiana sand dunes with Cowles. Shelford had been Child's graduate student in 1903, had done ecological research at Friday Harbor in the state of Washington while Child was there, and later used many of his teacher's ideas in his ecological studies.[41]

If Child's conception of organismic pattern was compatible with an ecological approach, it was thoroughly in conflict with the new genetic analysis of inheritance. Child's goal of creating a genuine synthesis of morphology and physiology was at odds with the trend toward specialization that was going on elsewhere, a trend that was especially evident in the new work of Mendelian geneticists, who tried to construct chromosome maps from knowledge of how traits were inherited over several generations. In developing his theory of organismic unity, Child has assumed, along with most embryologists, that problems of development were connected to problems of heredity. Belief in the basic identity of these two processes was especially meaningful in a neo-Lamarckian context, if one thought of the organism as acquiring its character partly through adaptive response to the environment, whether as embryo or adult. Child's theory supported a view of heredity and evolution that eventually put him into direct opposition to Thomas Hunt Morgan, who after 1910 built an extremely successful experimental program by frankly adopting a morphological approach to genetics, when it became clear that a physiological approach was too complex to yield clear answers about gene function.[42]

Child reasoned that if the organism was a specific reaction system in which differences in metabolic rate initiated growth and development, the unit of inheritance had to be the organism itself. What was inherited was a basic reaction system, not a group of chemicals or a group of characters as distinct entities.[43] Child denied the distinction between soma and germ plasm established first by German zoologist August Weismann and later reinforced by Morgan.[44] He denied that heredity could be localized in any specific part of the cell. Moreover his theory, in opposition to Weismann's and Morgan's, was compatible with a Lamarckian view of evolution, for if the organism as a whole was the unit of inheritance, acquired characters might be impressed on the organism by the environment and might over several thousands of generations become permanent. As he explained, "If the organism is in any sense a dynamic entity, then its evolution must be a reaction determined, on the one hand, by its physico-chemical constitution, and on the other, by its relations with the external world, and its adaptations are simply special features of this relation."[45]

This hostility toward genetics was characteristic of the Department of Zoology at Chicago, which was dominated by embryologists. When Sewall Wright arrived in the department in 1926 to teach physiological genetics, Child was openly skeptical about his chances of developing this field.[46] Embryologists remained suspicious of geneticists' incursions into embryological territory even into the 1930s.[47] In the 1910s Child's skepticism that genetics would lead to a genuinely synthetic physiological and morphological explanation of biological processes was well founded. In his hostility to the narrowness of genetics, even into the 1920s and later, Child was on common ground not only with fellow embryologists but also with naturalists such as William Morton Wheeler and Francis B. Sumner.[48]

Child was perhaps exceptionally rigid in dismissing Morgan's work as little improvement over Weismann's speculations about genetic mechanisms: he regarded both theories as equivalent to vitalism in their inability to explain the origin of pattern. Child's conservatism was no doubt reinforced by the fact that he had already worked out an alternative view of organismic pattern by the time Morgan made his celebrated conversion to the chromosome theory in 1910. But we should not overlook an extenuating circumstance that might have affected his judgment: the fact that certain socialists had embraced Weismann's views as an alternative to Lamarckian evolution. Arthur M. Lewis, for instance, delivered a lecture series to enthusiastic audiences in Chicago in the winter of 1907–1908, which surveyed various evolutionary theories and assessed their value for socialist reform measures.[49] Lewis made it clear that in the debates in the nineteenth century between Weismann and Herbert Spencer over Lamarckian evolution, Weismann's arguments against Lamarckian inheritance not only were better supported than Spencer's defense of Lamarck but were more adaptable to the socialist agenda. Lewis's book on social evolution, consisting of the first ten lectures in the series, appeared soon after the lecture series. Child, who leaned more toward a Spencerian view of biological and social evolution, would likely not have missed this connection between the chromosome theory and proselytism for socialism, and perhaps this link strengthened his hostility to genetics later.

The Ecology of the Social Organism

Child's interest in organism-environment relations and in the development of the nervous system as a central problem of biology raised the broader question of the biological basis of behavior, a subject he took up in the 1920s and published in 1924 in *Physiological Foundations of Behavior*. Here Child developed and extended his gradient theory in an evolutionary context, stressing throughout the book that the individual's dynamic interaction with its environment, through all stages of development, was crucial to an understanding of each characteristic. The gradient concept was important also for evolutionary theory, for the dominant regions served as pacemakers of function and development, and hence also of evolutionary development. Since in higher organisms the dominant organ was the brain, behavior could initiate evolutionary change.

What was most novel in this book was his extension of this physiological theory to explain social integration in the higher organisms, including human beings. Child had hinted in 1915 that one could extend the analysis of the biological laws of organismic unity to the human species but had not developed his ideas in detail. He was aware that metaphors were commonly drawn between the individual and the state:[50] this analogy was not fanciful, he thought, for a mechanism of control or government was essential to both.[51] In 1915 Child declined to elaborate on this analogy, but in 1924 he boldly marched into the territory of the sociologist, arguing that "if man is a product of evolution, the foundations of human society lie, not in the human race, but in other organisms."[52]

Child believed that there was a basic similarity in the general laws of cells and simple organisms at one extreme of the evolutionary scale, and the modern nation or state at the other end. There was, he argued, a true physiological continuity "not simply from the physiological gradient to the fully developed organism, but to the dominance of the idea in intelligent social integration." Moreover, Child thought that physiological evolution followed a general (though not inevitable) progressive course from autocracy to democracy. "The progress of [physiological] evolution then has apparently not been toward a socialistic or communistic form of integration. . . . The course of evolution of physiological integration has been in the main toward democracy with representative government, assisted so to speak by experts, the organs of special sense."[53] The evolution of the individual followed the same course as social evolution, a gradual move toward democracy.

The organism was a kind of institution resembling the state, which governed by means of relationships of dominance and subordination. The relation of dominance and subordination was a basic factor in all kinds of organic development, whether physiological or social. Child's argument is difficult to follow, but he seemed to reason backward from society to the organism. That is, he assumed that democracy was the ideal social organization and from that assumption inferred that individual development tended toward a parallel democratic form. Extending this analogy, he described the cerebral cortex of higher vertebrates as "a sort of physiological parliament": "Through lower centers it may receive reports from any part of the organism and its final action, the efferent impulse, depends upon these and upon its records of past events, i.e., memory."[54] Why a highly developed nervous system should represent a more "democratic" form of organization is not logically obvious, and in fact his argument is circular, but the analogy illustrates the strength of his conviction that social problems were biological in origin.

Just as in physiological evolution, where the dominance relations originated in reaction to an environmental stimulus, so also in social evolution these relations were historical products of interaction with the environment, for each person was both a product of his or her present and past relations to the environment, as well as a product of his or her longer evolutionary past.[55] Inheritance and environment acted together in all phases of evolution.

Child's theory suggested that change in human society depended largely on control of the environment, and that therefore real progress was possible despite the limitations that our biological inheritance might set on our ability to change. Because behavior was the pacemaker of evolution, and behavior in humans was often guided by ideas about the future, social control might take the form of making the future play a more important part in behavior. The point of view therefore was teleological in its emphasis on purposive action as an evolutionary force.[56]

His complete theory pointed toward an integration of biology, psychology, and sociology within a progressive evolutionary worldview that placed at the pinnacle of development the institutions and values of Anglo-American culture. His view of the individual was meant to correct the hereditarian position adopted by geneticists, for whom environmental effects were downplayed in their efforts to abolish once and for all the Lamarckian hypothesis, over which they had wrangled for many years. As I have suggested, Child may also have been reacting against the association between Weismann's theory of heredity and socialism. Child wanted to give the environment prominence in a way that did not raise the Lamarckian specter but still supported American progressive democratic values.

Child's theories did strike a responsive chord among social scientists, as he intended. His concept of the physiological gradient had notable impact on sociologists who, led by Robert E. Park and Ernest W. Burgess at Chicago, developed the field of human ecology between the wars. Human ecology involved the use of ecological concepts of the biotic community, its structure and development, in the urban setting in an effort to create an objective sociology. Human ecologists of this school drew heavily on community studies by plant ecologists and they also used Child's concept of the gradient to depict the business district as the "center of dominance" of the urban community, a center which exerted a pull or influence on other areas depending on their distance from the dominant center.[57]

Child's ideas also influenced contemporary biologists beyond the University of Chicago, in particular William E. Ritter (director of the new Scripps Institution for Biological Research in California) and English biologist Edward S. Russell, both of whom where hostile to genetics and reductionist biology and appreciated Child's holistic message.[58] Biologists studying animal behavior incorporated Child's ideas into their work as well. Clarence R. Carpenter, who studied the social behavior of primates under Robert M. Yerkes at Yale University in the 1930s, referred to Child's work on the physiological gradient in the context of his research on dominance hierarchies in primate groups.[59]

A more direct and pervasive impact of Child's later thought, however, can be seen in the work of his colleague in the Department of Anatomy, Charles Judson Herrick, who was interested in psychobiology and developed his ideas within a progressive, evolutionary perspective, expounding a view of human social evolution very similar to that of Child.[60] Child and Herrick drew constantly on each other's ideas in the 1920s, and part of their writings represented a genuinely collaborative effort. What drew them together was a common physiological approach to the study of behavior, a general emphasis on organism-environment relations, and a willingness to extend biological observations to human society within a progressive evolutionary context. In the next section I shall consider Herrick's psychobiological program especially in relation to Child's ideas and as a representative product of the "Chicago school."

The Natural History of the Human Psyche

Charles Judson Herrick was born in Minneapolis in 1868, the son of an itinerant Baptist preacher. He intended to become a minister but decided in college that he was not suited to the clerical life and instead began to study biology. In 1891 he obtained a bachelor's degree from the University of Cincinnati, where his brother Clarence Luther Herrick (1858–1904) was teaching in the Department of Natural History. In 1892 Clarence was hired by President Harper of the University of Chicago for the purpose, he believed, of setting up an interdisciplinary program in neurobiology and comparative psychology that would cut across departmental boundaries. His grand objective was a survey of the original nature of man, man's evolutionary origins, and his ecological relationships in all their aspects, including the physical and the social environment. Such a program implied cooperative work in several fields and disciplines that would normally be represented by different departments—departments usually separated, as his brother Charles remarked, by "rigid and jealously guarded barriers;"[61] these included anatomy, physiology, zoology, ecology, anthropology, psychiatry, the social sciences, and philosophy. As it turned out, the actual position given to Herrick allowed him far less scope to develop such a cooperative program, and he resigned in bitterness even before he went to Chicago.[62] Clarence was then hired by Denison University in Ohio, and Charles followed him there as a graduate student and later member of the faculty. In 1897 Clarence Herrick became the second president of the University of New Mexico; he resigned in 1901 and died of tuberculosis if 1904. Charles Herrick eventually helped to fulfill Clarence's dream of creating a new field of psychobiology.[63]

Charles Herrick had taken a leave of absence from Denison in 1896 to do doctoral research at Columbia University under Henry Fairfield Osborn and Oliver S. Strong. His thesis was a functional analysis of the nerves of the head and spinal cord in a species of minnow. The degree was awarded in 1900, and in 1907 Herrick accepted a professorship in neurology in the Department of Anatomy at the University of Chicago. He became emeritus professor in 1934 and remained at Chicago until 1937.[64] In 1942 he was named visiting professor emeritus at the University of Michigan.

Even late in his career, Herrick insisted that the point adopted by both himself and his brother was that of the naturalist, despite the fact that he worked in the laboratory. By "naturalist," he meant a biologist who took care to know the whole life history of the species under examination. Herrick recognized that there had been a general move from field natural history to laboratory science in the late nineteenth century and that his brother's career, starting with natural history survey work and ending in the physiological laboratory, was a typical expression of that shift. But he pointed out that Clarence's enthusiasm for field natural history never weakened even as his interests changed to physiological problems.[65] Like his brother, Charles Herrick also retained the naturalist's desire to know his experimental subject completely, its anatomy and physiology, its life history and evolutionary history. The naturalist's sensibility found expression in Herrick's constant pursuit of an interdisciplinary perspective, viewing the problems of biology and psychology, of human beings and the lower animals, as inseparable. Ultimately Herrick was trying to accomplish two things: first to develop comparative neurology as a pure science, not as a handmaiden to medicine, and second to promote a synthetic science of psychobiology which explored human life in all its aspects.

The starting point for the synthesis was to think in dynamic terms, that is, to recognize that structure could only be understood in relation to function. In 1904 Herrick expressed what he considered to be his brother's principles of dynamic science as applied to morphology, a science consisting not merely of the description of form but of the explanation of form, meaning the relation of form to function and to evolutionary history. In effect, Herrick was arguing that morphology was not properly a field unto itself but had to reach out into the other branches of science. Applied to neurology, the dynamic approach means recognizing that the key to neurological structure was behavior, or, as Clarence phrased it, "structure is behavior in instantaneous photograph."[66]

The roots of this synthetic approach were in both brothers religious as well as scientific: science, along with religion, would provide a guide to human behavior and conduct, in a way that vindicated Christian belief and democratic forms of government. Clarence Herrick was deeply religious, having been raised in the home of a Baptist minister. His interests in psychology and philosophy, and in the study of evolution, development, and the function of the nervous system, could be seen as a natural outgrowth of his religious background.[67] Indeed Clarence brought psychology into the chapel, literally as well as figuratively, for he actually used the chapel at New Mexico for lectures on anatomy and embryology.[68]

Herrick's research was part of that tradition, which Robert J. Richards has analyzed, growing out of Darwin's theory of evolution and culminating in a large scientific literature on mind and behavior in the late nineteenth century.[69] The work of George J. Romanes and Conwy Lloyd Morgan in

Britain and the psychological studies of William James and James Mark Baldwin in the United States, created a vigorous research tradition that focused on the origins of human social behavior. Richards argues that biological theories about mind and behavior were proposed in an effort to resolve spiritual dilemmas raised by Darwin's theory. Therefore to understand how psychology and behavioral biology developed, one must also understand the spiritual crises experienced by the leaders in these fields.

Charles Herrick's interests in biology and psychology may have been motivated not so much by a concern with the spiritual dilemmas raised by Darwinism as by the converse, namely his dissatisfaction with the intolerance toward open inquiry into the nature of spirituality that he encountered in his early theological course.[70] But the transition from theology to biology did not entail a serious intellectual crisis, nor a rejection of his belief in the importance of the spiritual life. By transferring the study of spirituality from the theological to the biological plane, however, Herrick was challenging the authority of the religious leader. He rejected the idea that there was a barrier between the spiritual and the physical, an idea that placed the spiritual life outside the domain of science. Instead, Herrick envisioned the naturalist and human biologist as working to refine and clarify spiritual values.[71]

In psychobiology, science, philosophy, and religion converged, but in a way that strengthened the scientist's authority over that of the priest. Herrick would later argue that moral values had superior survival value in a biological sense, because they encouraged the kind of altruistic behavior necessary for a stable, progressive society.[72] A similar defense of the innately moral nature of man could be found in Darwin's *Descent of Man*. Darwinian science once and for all defeated the doctrine of original sin. Herrick's religious views were in keeping with the liberal interpretation of the function of religion in a social context that was associated in particular with the Chicago Baptists.

The dynamic approach to psychobiology was also in keeping with the philosophy of pragmatism as developed by Charles S. Peirce, William James, and John Dewey. This connection was not lost on biologists like Clarence Herrick, who pointed out that the conviction of the harmony and the unity of the psychological and the biological was the justification of pragmatism as a method, if not as explanation.[73] I shall explore the connections between pragmatism and psychobiology more fully in a later section.

Clarence Herrick himself did not fully achieve his ideal of a dynamic biology; the first thirteen years of the *Journal of Comparative Neurology*, which he founded in 1891, were mainly devoted to structural and not functional aspects of neurology.[74] Charles hoped to realize the unattained goals of his brother in developing a genuinely functional analysis of the nervous system. At Chicago he found the financial support and time for research that he needed. His laboratory studies focused on the neural mechanisms of the vertebrate brain, his own speciality being amphibian neuroanatomy. The guiding principle of his research was to interpret structure in terms of function and to relate the functioning of the brain to overt behavior. He pursued these relationships through comparative study of the vertebrate brain. His classic analysis of the structure of the salamander brain and its functional interpretations, published in 1948, was his last major original work.[75] Herrick's teaching introduced a functional approach to neuroanatomy that departed from the normal gross anatomical course given to medical students at Chicago.[76] His ideas about neurology influenced medical teaching more broadly in the 1920s when his methods were picked up by medical textbook writers. This comparative method of neurological study contributed to what is known as the "American school of neurology."

But the main goal of his intensive labors in comparative anatomy was to help unravel the human nervous system and discover ways of using it more efficiently. This goal was well beyond the reach of a single investigator. Psychobiology entailed a multidisciplinary attack on all aspects of human physiology, behavior, and psychology: this rounded approach, which he called a natural history of the human race, would enable scientists finally to connect physiological and mental processes. Discussions of psychobiology at Chicago attracted an interdisciplinary group of faculty who formed a "Neurology Club" that later included clinical neurologists and psychiatrists as well as biologists and psychologists.[77]

Herrick was certainly not the exclusive proponent of a dynamic physiological approach to behavior. His contemporary Herbert Spencer Jennings, whose work Herrick admired, had also gravitated to the new experimental biology and become the main advocate in the early twentieth century of the experimental study of behavior in lower organisms. Briefly in the early twentieth century Jennings was on the editorial board of Herrick's *Journal of Comparative Neurology,* along with Robert Yerkes, until the *Journal of Animal Behavior* was organized in 1910, whereupon both left.[78] Jennings's 1906 monograph, in part an attempt to locate the origins of psychic behavior in lower organisms, was like Herrick's work in the tradition of neo-Darwinian research on mind and behavior.[79] His goal was to study these evolutionary relationships by describing behavior not as something reducible to law, as Jacques Loeb had tried to do in his tropism theory, but as it actually occurred, or, as he later said, "as a moving picture would show it."[80]

Charles Herrick arrived at Chicago in 1907 with a dynamic conception of the organism that had much in common with Jennings's views. Small wonder that he found so much to admire in Child's own dynamic biology or that the two, perhaps working out their ecological perspective together during long walks over the Indiana dunes, discovered they had the basis for a genuine collaboration.[81] The outcome of this collaboration was the enunciation of a concept of human nature that reflected the intellectual environment of the University of Chicago and expressed a distinctively liberal response to the postwar culture of the 1920s, an example of the "forward-looking liberalism" that pundit H. L. Mencken so enjoyed debunking in these years.

The Physiology of Liberalism

The collaboration of Child and Herrick found expression in Herrick's book, *Neurological Foundations of Animal Behavior* (1924), which was the counterpart to Child's *Physiological Foundations of Behavior,* published the same year. Herrick was searching in this book for the biological origins of behavior, with an eye on the human species, for even man's most artfully contrived mechanical and social fabrications "may strike their roots down deep in to the biological soil from which the human race as a whole has sprung."[82] His perspective was evolutionary in the spirit of Darwin and Spencer, though he would not commit himself to a wholehearted acceptance of natural selection as the main force in evolutionary change.

Herrick's indebtedness to Child was evident on nearly every page. He used Child's theory of the relation between protoplasm and environment, arguing that behavior had to be seen from a physiological point of view, focusing on the reactions of the organism to external influences. He accepted as adequately demonstrated the principles of protoplasmic transmission of excitation, of physiological gradients, and dominance and subordination of parts. His book extended these ideas to the neurology and behavior of the higher animals. Even in the human species, behavior could be understood in terms of the correlation between regions of higher dominance (the cerebral cortex), which reinforced, inhibited, or redirected the nervous currents of the lower centers.[83] Nervous systems of higher animals might grow and act according to biological laws different from polyps and protozoans, but these laws had a common origin and were of the same basic type. Like Child, Herrick was antivitalistic in his interest in organismic unity, arguing that unity involved no mystic principle, no immanent agent or force within the body.

One of Herrick's major theses tied in with the work of William Morton Wheeler on the social organization of ants. Wheeler had pointed out that all species possess both innate, stable forms of behavior as well as more plastic and modifiable forms. Each type of behavior had to be recognized separately. It did not make sense, therefore, to express the whole of life, including the human mind, in terms of successive hierarchies of reflex or other determined modes of behavior. Herrick emphasized that humans could adjust their behavior in ways that were not inbred but were the product of individual contact with the world, and perhaps of conflict with it.[84]

Herrick argued that intelligence was creative in a literal sense; intelligence had the power to shape character. It was the constructive power of intelligence that made mental and other kinds of progress in humans possible. The mind was a causative agent in evolution, for conscious processes of the mind shaped human conduct and the thinking mind (or functioning brain) affected other organs of the body. Like Child, he considered behavior to be the pacemaker of evolution. Because he believed in the active power of the mind, Herrick distanced himself from the standpoint of the radically objective behaviorist school, which disparaged the value of introspective experience.[85] Herrick, while otherwise considering himself to be a behaviorist because of his biological approach to the mind, wanted to preserve the value of introspection as a psychological method.

In his views of future progress, Herrick retained an optimistic belief in orthogenesis, or progressive evolution, as did his former teacher H. F. Osborn. It was impossible to tell whether real physical evolution was still in process, but certainly social evolution was still occurring and all forms of evolution were connected: there was no radical separation between physical and cultural evolution. Herrick interpreted Child's evolutionary views as implying that such characteristics as commercial interchange, patriotism, and altruism were as truly organismic factors as were the physiological gradients of the flatworm. Social evolution was a problem to be tackled through biology.[86]

Although these theories of social evolution were tied to a long Darwinian tradition, the prominence of these arguments after the First World War points to the effect of the war in raising biologists' consciousness of these issues. Gregg Mitman argues that biologists were stimulated by wartime rhetoric suggesting that German militarism had been buttressed by Darwinian views of nature red in tooth and claw.[87] In response they asked whether an alternative American view could not be devised, which would promote stability and gradual progress. In working out their ideas about an alternative "sociobiology" emphasizing the biological basis of cooperation, they began to make explicit the connection between biology and social theory. Although the views they advanced originated in the Darwinian tradition that Richards discusses, the war forced to the surface ideas about the application of biology to society that were only vaguely developed in their earlier writings. Moreover, after the war Americans were faced with a growing climate of violence, revealed by public hysteria against Bolshevism and the threat of class warfare, and by heightened labor and racial tensions. In July 1919 Chicago erupted in race riots, fueled by gang warfare, which continued for five days.[88]

The relationship between the sociobiological theories of the 1920s and the wartime and postwar experience of biologists needs to be explored in more detail. I have no direct evidence to document how Child's ideas may have been related to neo-Darwinian or other arguments made during the war, though he was probably aware of these debates. In Herrick's case it is clear from his later comments that he was alarmed at how easily people might be led into senseless aggression by wartime propaganda. He began in the 1920s to formulate a view of social progress predicated on finding a balance between individual assertiveness and cooperative activity.

The success of modern society, he believed, depended on harmonious interaction between various parts of society, under the guidance of dominant groups. These groups dominated because of their greater dynamic efficiency. The power exercised by the dominant groups was not tyrannical, but a natural regulatory control developed within the community. Herrick accepted the ideal of a meritocracy grounded in evolutionary biology: leaders dominated because they had superior vision and ability, and it was right for them to be given leadership positions. In a democracy whose values were supposedly recognized by each member of society, evolution would be progressive: "When once society has definitely set its face toward the higher standards of social relationship no single community can obstruct the general movement. Nor can any advanced people revert to the barbaric standard of isolated self-sufficiency. They must adjust to the general movement and advance with it or perish from off the earth."[89] In assessing human evolutionary progress, the right question was not whether the human mind had greater intellectual capacity than in the past. Rather, one needed to know whether the culture of today showed better teamwork than of old, whether the machinery of cooperative effort was more efficient: were people more neighborly? It was not the individual, but the social organism, that was the unit of evolutionary progress.

Earlier I alluded to the connection between the dynamic, evolutionary point of view and the philosophy of pragmatism. Now we can explore this link in more detail, for Herrick referred to this literature, and John Dewey's work in particular, in the context of his biological discussions of individuality and social evolution.[90] Pragmatism as formulated by Peirce, James, and Dewey was a response to traditional conceptions of mind, perception, and reason. Historically, philosophical conceptions of mind treated reason as the passive spectator of data provided by perception and/or introspection. The central problem of knowledge was understanding how (or if) reason derived universal truths from the data available to it. This conception of reason underlay the traditional distinction between science and art, or between theory and practice. Science was defined as the analysis of sensory and introspective data and the derivation of underlying universal laws. Practical arts, by contrast, were viewed as the subrational pursuit of particular ends through more or less mechanical procedures.

This distinction between theory and practice was challenged in the late nineteenth century when the hitherto humble arts of medicine, material science, and applied mathematics began to produce characteristically "theoretical" programs of research. Philip Pauly, for instance, discusses how the link between theory and practice was achieved in the philosophies of Ernst Mach and the Austrian engineer Josef Popper-Lynkeus, which in turn influenced the development of Jacques Loeb's ideas about science as engineering.[91] At this time physics, philosophy's model of science, also began to use statistical generalizations in place of universal natural laws. But whereas the distinction between theoretical science and practical art was breaking down by the end of the century, the conception of reason which underlay it had not yet been abandoned. The pragmatists united in attacking this conception.

The Chicago pragmatists, led by Dewey in the early twentieth century, argued that the rejection of the older conception of science as pure insight into the world entailed rejection of the conception of the mind as a passive spectator. Science stood revealed as an activity with the same goal as the explicitly practical arts, control of and adaptation to an environment. Intelligence was just the name for that activity abstracted from its particular applications. Intelligence was therefore a causal factor in the production and form of an individual's experience.

Dewey's philosophy was premised on an ecological view of the individual as evolving in constant interaction with its environment. The pragmatist point of view always emphasized the complexity of these interactions, resisting the temptation to "pulverize" the individual into isolated sensory qualities or simple ideas. The individual could never be isolated from the physical and social environment. The individual, moreover, had the capacity to shape the future through intelligent action, to bring a more desirable future into existence. In Dewey's words, "A pragmatic intelligence is a creative intelligence, not a routine mechanic."[92] Although the general point of view was progressive, it was not meant to be teleological or blindly optimistic: progress was not necessary or automatic.

Dewey's pragmatism was connected explicitly to a democratic view of philosophy. Knowledge and philosophy were not the possessions of an elite class of learned specialists. Knowledge derived from any kind of creative use of human intelligence. Dewey hoped that this philosophy would prove relevant to America's needs and, in the context of the First World War, would steer America clear of the dangerous deification of power and the nationalism that he perceived to be inherent in German philosophy. He was concerned that philosophy should not help to justify wartime aggression as a moral necessity, as it appeared to do in Germany. He also wanted to use philosophy to find solutions to social problems. The pragmatist was a social engineer.

American society in the interwar years seemed to be characterized by basic ambiguities, which Dewey identified as the source of modern discontent. These ambiguities were a materialism and narrow practicality, on the one hand, and an idealistic devotion to social service, on the other; an ethos of individualism, yet a constant pressure toward standardization and conformity; a desire for peace and respect for the rights of others, yet the pursuit of policies of economic imperialism.[93] The question facing the modern critic was to explain why and how these opposing pressures coexisted.

Dewey and many others saw these ambiguities as an apparent product of scientific and technological change in industrial society. The union of industry, commerce, and science that was under way in the twentieth century challenged the philosophic tradition that separated knowledge and practice. A dominant concern was how the individual should adjust to the needs of the new industrial society, and how this adjustment could be made while preserving the values and institutions of a democracy. For Dewey, one solution to the unrest of modern industrial civilization was to develop a pragmatic philosophy based on a realistic assessment of modern society and its future prospects.

The biologist, no less than the philosopher, was involved in this debate, for biological theories defined the nature of the individual and therefore had implications for the preservation of democracy. The expression of the liberal stance in biology that was the counterpart of Dewey's pragmatism focused on the idea of the mind as a creative force in evolution and the implications of that view. In the biological argument, intelligent behavior was the pacemaker of evolutionary change. This point of view stressed the complexity of the individual and therefore was holistic and antireductionist, though at the same time it insisted on adherence to an experimental method. The issue was whether the individual could act as a free agent, capable of designing the future through intelligent action.

Herrick saw the individual as a complex product of heredity and the environment. Although eugenic improvement was possible, selective breeding was a slow process. A stable society composed of altruistic individuals who would "refuse to be stampeded into war by hysterical propaganda or infantile ideals of national aggrandizement" could be created most efficiently through social control rather than eugenic breeding: social control meant instilling the right ideals through education.[94] Herrick considered this kind of social change to be the practical equivalent to a genuine change in human nature, to all intents and purposes as permanent as any change created by germinal evolution.

His definition of human nature as encompassing hereditary and environmental influences led him to use rhetorically the concept of "emergent evolution," made prominent in the 1920s through the writings of Conwy Lloyd Morgan and William M. Wheeler.[95] This doctrine rejected the idea that the universe could be reduced to matter and motion. Instead, the idea of emergence was that at each level of organization the system as a whole acquired new features because the constituent parts themselves acquired new properties, new modes of action, by becoming part of the system. The doctrine was not vitalistic, for experiment remained the main method of science and the only method of discovery. But in the organized system the parts could not be analysed as though they were isolated.

The theory was a variant of the "organismic" point of view that many biologists had adopted to distinguish biology from the physical-chemical sciences. But in these discussions in the 1920s there was another agenda, a political one, behind the scientific and philosophical discourses. Jennings had succinctly linked the doctrine of emergent evolution to liberal ideology in an address delivered to the American Association for the Advancement of Science at the end of 1926.[96] I shall digress briefly to review Jennings's arguments, for they articulated especially clearly the connection between emergent evolution and views of human progress, within the context of his belief in creative intelligence. Although Herrick's views were less well developed in print, I believe that he thought the same way and that we can use Jennings's article to explore this common logic. It is highly likely that Herrick knew of Jennings's address, even when he did not cite it in support of his own arguments, and he may even have been influenced by Jennings's discussion. There were, however, some differences in their attitudes toward the social role of the scientist, to which I shall return.

Jennings was not interested in the doctrine of emergent evolution merely as a philosophical exercise. Its importance was precisely in its clear implications for practice on several levels. First, the strictly reductionist view of science placed experiment in a secondary role, because once the laws of matter were known, the behavior of the system became simply a matter of computation. In a deterministic science, experiment was needed only to test the accuracy of the computed results. But in the doctrine of emergent evolution, the properties of the whole could not be predicted from knowledge of the parts in isolation. They had to be *discovered* by experiment. Experiment was therefore the primary and final method of science, and the only method of discovery. The biologist, by adopting this perspective, broke free of the hold of the inorganic sciences on biology, so that the doctrine became an intellectual tool in the struggle for authority between biology and the physicochemical sciences, a struggle that in the 1920s was pushing physiology toward a more reductionist method: "The doctrine of emergent evolution is the Declaration of Independence for biological science."[97]

Within biology also, practical results followed from the principle of emergent evolution. For Jennings, the awareness of emergent properties and the diversity of the organic world would restrict the biologist's ability to pronounce on matters of human conduct. Any such pronouncements could only follow intensive study of the human species, not just from the biological side, but from the political, economic, and historical sides as well. Society showed unique emergent properties which could not be compared with those in the rest of the animal world. The doctrine of emergent evolution at once elevated the biological way of thinking so that biology could operate independently of the physical sciences, while restricting the biologist's claims to act as expert in human affairs. The doctrine thus also served as a "Declaration of Independence" for the social sciences. Jennings's remarks reveal an ambivalent attitude toward the idea that human progress can be controlled through biology, and an underlying ambivalence about the scientist's duty to step into political debates. He was wary of sullying the image of science, as had been done in the eugenics movement by overenthusiastic proponents of social engineering.

Jennings's criticism of reductionist biology was influenced also by implications he saw in reductionism for human progress. A strictly reductionist philosophy which did not take account of emergent properties led to the belief that mental activities—ideas, ideals, beliefs, purposes—had no place in human conduct. The reductionist viewpoint combined with the theory of natural selection—the doctrine of survival of the fittest—led to a belief in the inevitability of the destructive laws of nature. If one adopted this logic, war became a means of progress and the only justified conduct was the exertion of force. Reductionist science was forced into this conclusion because it could not allow a causal relation to exist between the mental and the physical. Admitting such a relation would destroy its ability to save the predictability of events, and "these it must save or itself perish in the struggle for doctrinal existence."

The emergent doctrine differed from reductionist science in giving the emergent properties status as causal agents: "Thought, purpose, ideals, conscience, do alter what happens." The same was true on the social as well as individual level: "The desires and aspirations of humanity are determiners in the operation of the universe on the same footing with physical determiners." Moreover, this doctrine, far from encouraging any doctrine that made the individual subservient to the whole, actually reaffirmed the independence of the individual as a unique product capable of acting in ways different from other individuals under the same external conditions: "Such an individual is free from the tyranny of general law; is free from determinism by conditions outside itself; is free to act in accordance with its own nature alone; and yet in its acts there is no breach of experimental determinism."[98]

Herrick used the idea of emergent evolution in essentially the same way as Jennings, though he did not work out the argument in the same detail. For Herrick, as for Jennings, emergent evolution and the holistic definition of the individual that it implied was not used to map out a specific research agenda in evolutionary biology. It served a broader rhetorical function in affirming the difference between biology and the physical sciences. Both Jennings and Herrick were against the procrustean reduction of biology to physics and chemistry. But the doctrine was at the same time a statement of a liberal political ideology, an ideology that permeated the scientific positions that these men adopted. The crux of the argument was to find a biological justification for individual freedom and for the ability of humans to shape their future through intelligent action.

Their efforts to define this ideology in a scientific context appear to have been stimulated by the social problems of the postwar era and by the arguments marshaled to justify aggression during the war. Competitive social Darwinism and reductionist approaches to biology were thought to be related. In response, Jennings and Herrick advocated a biology that gave primacy to individual intelligence as a creative force in evolution, and that was sensitive to the complexity of historical circumstances that shaped the individual.

Science and Ideology

The holistic biology of Child and Herrick suggested a very broad definition of animal and human "nature," a nature that was not reducible to a set of genes or chromosomes passing from one generation to the next. Their viewpoint emphasized the totality of the individual in its environmental and social context. Although their ideas were framed in the language of modern experimental science, their perspective reflected the sensibility of the naturalist, for whom biology meant the study of the whole organism in all its aspects. This was how Whitman, who built the biological program at Chicago and shaped the direction of American biology in the 1890s, understood biology.[99]

From time to time between the wars, those trying to preserve the naturalist's definition of biology inveighed against reductionist methods. The persistence of the naturalist's definition of biology was signaled by the various appeals to holistic concepts made by members of Child's and Herrick's generation in the 1920s. The younger generation regarded these appeals as at times blockheadedly resistant to new directions in science. Indeed, holism was largely nostalgic, though its message was not antiscientific. Francis Sumner took the opportunity of a symposium in 1939 on "The Biological Basis of Social Problems" to stress the value of the synthetic methods of the naturalist, who "never forgets that he is dealing with organisms dwelling in environments" even when talking about genes, ions, or colloids.[100] He saw the specialization of science as an antisocial move, against the true spirit of science that looks to the larger whole. Child and Herrick would have sympathized with his dislike of specialization and the changes that biology had undergone since the turn of the century. In developing a version of human biology as a kind of natural history of the human race, Child and Herrick were also preserving the naturalist's sensibility in modern experimental science.

The biological examples I have chosen reflected the same ideology and the same enthusiasm for scientific method and practical application of science that was expressed in American pragmatism. Both science and philosophy were responses to the demands that modern industrialization made on the individual: as Bertrand Russell aptly remarked, pragmatism was the philosophy most appropriate to industrialism.[101] And in linking their concerns with those of the pragmatists, these biologists created a niche for themselves as philosophers—not philosophers divorced from practical concerns or enmeshed in mysticism, but biological pragmatists contributing to the solution of modern problems. In the 1920s these debates took on a political dimension, and biology became the ground for discussions of the conditions of stability in modern democracies. It may be that this connection with the pragmatic tradition and the political discourse that was associated with it was what gave American biology, and American holism, its special "American" character, distinct from the European and British traditions.[102]

Whereas Jennings was ambivalent about the role of the scientist in social debates, Herrick had no qualms in declaring that a fundamental problem of science was to understand human motivations in order to control social movements and human destiny.[103] He preferred social control through education, which he thought should produce individuals who were prepared to tackle problems in a spirit of tolerant cooperation. What made a creative, cooperative effort possible was the level of material wealth generated by science in an industrial society. Science therefore served two purposes: by revealing the causes of human behavior science enabled the creation of a society of individuals who "freely" chose to behave according to a given moral standard, and it created the material conditions that rendered the competitive "struggle for existence" ethos of laissez-faire liberalism unnecessary. Herrick echoed a contemporary argument that signaled a general ideological shift within both industry and science after the war, a shift away from the laissez-faire emphasis on competition toward a "corporate liberalism" which was more receptive to cooperation and planning.[104]

Herrick's reform goals were only vaguely developed, reflecting a general aversion in pragmatist philosophy to any kind of system making. As Dewey had asserted, the method of pragmatism was not teleological: there were no fixed ends. The crucial thing was the method itself: "Scientific method would teach us to break up, to inquire definitely and with particularity, to seek solutions in the terms of concrete problems as they arise."[105] The idea of an all-embracing system seemed to contradict the idea of developing an integrated individuality. Despite the vagueness of these reform measures, however, social control for Herrick was meant to produce permanent changes in society.

With the onset of the Depression in the 1930s the need for reform became more pressing, and by 1938 Herrick reflected that radical changes in social, economic, and political organization were long overdue in the United States.[106] He did not consider capitalism to be a panacea. Yet reforms had to evolve gradually on a local level, not be forced by the state. From Herrick's perspective, the political alternatives were too extreme—fascism on the one hand and Soviet totalitarianism on the other—leaving him no choice but to reject revolutionary change as less efficient than slower, evolutionary change. Politically a Republican, he also was wary of the kind of government intervention exemplified in the New Deal. Even his notion of educational reform, the key to all social change, stressed that education should be controlled not by the state but by the local community.

This wariness of bureaucratic control may have reflected Herrick's Baptist upbringing, for the maintenance of local autonomy was very important in the Baptist faith. Baptist churches began to band together in the United States in the eighteenth century, but these were cooperative groups that did not enforce doctrines or practices on the congregations within the country.[107] Among the central Baptist principles was the belief in the autonomy of the local church and the democracy of cooperation among the churches. In fact the Baptists were the last of the major denominations in the United States to have a single coordinating body. The Northern Baptist Convention, formed in 1907, represented their first effort to centralize, but the convention lacked the power to coerce the member churches. One would expect a person with Herrick's background to be especially sensitive to the need to preserve local autonomy within a large union.

In discussing social problems, Herrick took it for granted that science was a method of engineering. William Graebner has discussed the rise of what he terms a method of "democratic social engineering" between the wars, a method designed to promote cooperative behavior within the context of small-group activities (in church, school, or clubs) where nonauthoritarian, democratic values would be instilled.[108] He sees the widespread interest in group decision making as a method of teaching democratic values to be a typical middle-class response to the uncertainties of postwar society. The articulation of an ideology and method of social engineering within the fields of religion, education, and social science was part of a general tendency to elevate the ideal of "democracy" to almost religious status in this period. Herrick's applications of biology to social problems were one aspect of this move to preserve democratic values through a subtle, nonauthoritarian form of social engineering that rested on a holistic biology.

Another kind of engineering approach is reductionist: it seeks to control the organism by focusing on those aspects of behavior that can be predicted precisely. Loeb's science epitomized this engineering ideal in the early twentieth century. His example stimulated his younger colleagues to create new approaches to biological engineering ranging from radical behaviorism to the birth control pill.[109] But one need not be a reductionist to advance an engineering approach to life. Herrick may have disagreed with the behaviorist's methods because they did not include the study of introspective experience, and he may have appealed to the sensibility of the natural historian in arguing that one should know the organism in its entirety, but his goals were no less those of the engineer. In this holistic perspective, the individual was not dehumanized, however. Rather, the object of the analysis was to show that social control was the corollary to self-control and self-determination. Ironically, Loeb, disgusted by the First World War, began to retreat from his engineering standpoint toward a purer and more detached image of science just as Herrick and others were articulating their own engineering views in response to the war.

The Second World War made Herrick believe even more strongly that the biologist could not refrain from involvement in human affairs through fear of losing "objectivity." In his last book, published in 1956, which summed up his life's work, he described a new field that he saw developing, devoted to the search for the biological origins and nature of human patterns of social organization.[110] He called this new field "sociobiology," a term that had been in use since the 1930s. Herrick's sociobiology was not well defined as a field; he merely pointed toward its possibility but did not attempt the larger synthesis himself. Modern sociobiology owes less to Herrick than to the synthesis of population genetics and population ecology that occurred in the 1960s. Nevertheless, these earlier efforts to define a human biology between the wars should be seen as part of the history of modern sociobiology for they are part of the same evolving debate about the biological basis of human nature.

Human biology from a nongenetic point of view implies a different definition of human nature from that of the modern sociobiologist. When applied to humans in fields such as anthropology, modern sociobiology becomes a science of human motives, a science postulating that a large part of human behavior can be explained in the light of evolutionary strategies for maintaining biological fitness. Such explanations try to rationalize behavior in neo-Darwinian terms and therefore play down the role of fortuitous historical events or purely cultural factors as explanations of social institutions. To the extent that the sociobiologist is usurping the role of the social scientist, the science is potentially at least a tool for engineering society, despite the disclaimers of its adherents.

The significance of modern sociobiology, and the reason we should analyze its arguments carefully, lies in the way it has the potential for advancing an engineering standpoint within the guise of "objective" and orthodox neo-Darwinian biology, a biology often interpreted in a reductionist framework. For unlike Herrick's science, which embraced the whole individual as a historical product of environment and inheritance, modern sociobiology interprets behavior mainly from the point of view of the gene (though a good sociobiologist would acknowledge the importance of cultural evolution also).[111] Thus there arises a science that has profound implications for human social behavior and social planning, yet which in an important way is in danger of losing sight of the human being, of ceasing to be a "natural history" of humans in Herrick's sense. Herrick's science, for all its faults and vagueness, did not embrace this reductionist view of human nature, whose effect, one fears, is to deprive humans of control over themselves by neglecting their "whole" natures. What this definition means for the modern social engineer and how its dangers might be avoided are still open questions.[112]

Acknowledgments

I am grateful to the students in my graduate seminar of 1988 for their responses to an earlier version of this paper, and to Jennifer Welchman for helping me to understand John Dewey's philosophy. Research was in part supported by National Science Foundation grant SES 86-08377.

Notes

1. Daniel J. Kevles, *In the Name of Eugenics: Genetics and the Uses of Human Heredity* (New York: Knopf, 1985).

2. A survey of different approaches to human biology is in Edmund V. Cowdry, ed., *Human Biology and Racial Welfare* (New York: Paul B. Hoeber, 1930).

3. Stephen J. Cross and William R. Albury, "Walter B. Cannon, L. J. Henderson and the Organic Analogy between the Wars," *Osiris,* 1987, *3:* 165–192.

4. Sharon Kingsland, "Raymond Pearl: On the Frontier in the 1920s," *Human Biology,* 1984, *56:* 1–18.

5. Donna Haraway, "The Biological Enterprise: Sex, Mind, and Profit from Human Engineering to Sociobiology," *Radical History Review,* 1979, *20:* 206–237; idem, "Signs of Dominance: From a Physiology to a Cybernetics of Primate Society, C. R. Carpenter, 1930–1970," *Studies in History of Biology,* 1983, *6:* 129–219.

6. On the history of experimental psychology, see Robert Boakes, *From Darwin to Behaviourism: Psychology and the Minds of Animals* (Cambridge: Cambridge University Press, 1984).

7. Gregg Mitman, "Evolution by Cooperation: Ecology, Ethics, and the Chicago School, 1910–1950" (Ph.D. dissertation, University of Wisconsin, 1988); idem, "From the Population to Society: The Cooperative Metaphors of W. C. Allee and A. E. Emerson," *Jo. Hist. Biol.,* 1988, *21:* 173–194.

8. Sharon E. Kingsland, *Modeling Nature: Episodes in the History of Population Ecology* (Chicago: University of Chicago Press, 1985), chap. 2.

9. Edward O. Wilson, *Sociobiology: The New Synthesis* (Cambridge, Mass.: Harvard University Press, 1975).

10. Hamilton Cravens, *The Triumph of Evolution: American Scientists and the*

Heredity-Environment Controversy, 1900–1940 (Philadelphia: University of Pennsylvania Press, 1978). See esp. chap. 4.

11. The spirit of rapprochement between biology and the social sciences at Chicago is captured in Robert Redfield, ed., *Levels of Integration in Biological and Social Systems* (Lancaster, Penn.: Jaques Cattell Press, 1942).

12. Mitman, "Evolution by Cooperation," n. 7.

13. Garland E. Allen, "Naturalists and Experimentalists: The Genotype and the Phenotype," *Studies in History of Biology,* 1979, *3:* 179–209.

14. See the perceptive study of the careers of Stephen Forbes and Jacob Reighard by Stephen Bocking, "The Origins of Aquatic Ecological Research in the Great Lakes Region" (M.A. thesis, University of Toronto, 1987).

15. Garland Allen meant roughly the same thing when he used the term "holistic materialism" as opposed to what he called "mechanistic materialism" in *Life Science in the Twentieth Century* (Cambridge: Cambridge University Press, 1978).

16. Roll-Hansen argues that holistic biology was merely a "philosophical" position rather than a genuine scientific position. I argue to the contrary that a holistic approach does influence what kind of research is pursued. But holistic biology in these examples is also experimental, like any mechanistic science, and therefore cannot be distinguished from alternative mechanistic strategies according to whether it is rigorously experimental or not, which is Roll-Hansen's claim. See Nils Roll-Hansen, "E. S. Russell and J. H. Woodger: The Failure of Two Twentieth-Century Opponents of Mechanistic Biology," *J. Hist. Biol.,* 1984, *17:* 399–428.

17. A good contemporary definition of liberalism is by John Dewey, *Liberalism and Social Action* (New York: Putnam, 1935). See also John Herman Randall, "Liberalism as Faith in Intelligence," *Journal of Philosophy,* 1935, *32:* 253–264. On liberal debate within the Chicago school of sociology in this period, see Dennis Smith, *The Chicago School: A Liberal Critique of Capitalism* (New York: St. Martin's Press, 1988).

18. Libbie H. Hyman, "Charles Manning Child, 1869–1954," *Biographical Memoirs of the National Academy of Sciences of U.S.A.,* 1957, *30:* 73–103.

19. Hilde Hein, "The Endurance of the Mechanism-Vitalism Controversy," *J. Hist. Biol.,* 1975, *5:* 159–188.

20. Jane Maienschein, "Heredity/Development in the United States, circa 1900," *History and Philosophy of the Life Sciences,* 1987, *9:* 79–93.

21. These two categories do not do justice to the variety or subtlety of biological approaches to the problem of organismic unity. For a discussion of the Kantian tradition of teleo-mechanism in the early nineteenth century, before the rise of the dominant reductionist school of the Berlin physiologists, see Timothy Lenoir, *The Strategy of Life: Teleology and Mechanics in Nineteenth Century German Biology* (Dordrecht, Holland: Reidel, 1982).

22. Frederick Churchill, "From Machine-Theory to Entelechy: Two Studies in Developmental Teleology," *J. Hist. Biol.,* 1969, *2:* 165–185.

23. Hans Driesch, *The Science and Philosophy of the Organism,* 2 vols. (London: Adam and Charles Black, 1908), p. 283.

24. Child used the term "individual" to refer not just to the whole organism but to any part displaying organic unity. The segments of the earthworm, for instance, could be thought of as "subordinate individuals." See C. M. Child, *Individuality in Organisms* (Chicago: University of Chicago Press, 1915).

25. C. M. Child, "The Regulatory Processes in Organisms," *Journal of Morphology,* 1911, *22:* 171–222. Child attributed the origin of this metaphor to European evolutionary biologists Eugenio Rignano and Yves Delage. One is also reminded in this metaphor of the "stream of consciousness " imagery developed by William James to describe mental processes.

26. C. M. Child, *Senescence and Rejuvenescence* (Chicago: University of Chicago Press, 1915), pp. 27–28.

27. C. M. Child, *Individuality in Organisms* (Chicago: University of Chicago Press, 1915).

28. Ibid., p. 38 and p. 41.

29. Ibid., pp. 171–175.

30. C. M. Child, *The Origin and Development of the Nervous System, From a Physiological Viewpoint* (Chicago: University of Chicago Press, 1921).

31. Jacques Loeb, *The Organism as a Whole, from a Physicochemical Viewpoint* (New York: Putnam, 1916).

32. C. M. Child, review of Loeb, *The Organism as a Whole, Botanical Gazette,* 1918, *65:* 274–280.

33. The response to Child's ideas in embryology is discussed by Donna Haraway, *Crystals, Fabrics, and Fields: Metaphors of Organicism in Twentieth-Century Developmental Biology* (New Haven: Yale University Press, 1976).

34. Libbie H. Hyman, "Charles Manning Child," p. 80. The research program that grew from Child's theories is discussed in C. M. Child, *Patterns and Problems of Development* (Chicago: University of Chicago Press, 1941).

35. On neo-Lamarckism at the turn of the twentieth century, see Peter J. Bowler, *The Eclipse of Darwinism: Anti-Darwinian Evolution Theories in the Decades around 1900* (Baltimore: Johns Hopkins University Press, 1983). Child's vision of biology as a science was clearly indebted to Herbert Spencer's definition in *Principles of Biology* (first published in 1864 and often reprinted). See Herbert Spencer, *Principles of Biology* (New York: D. Appleton, 1874), vol. 1, chap. 5.

36. Philip J. Pauly, *Controlling Life: Jacques Loeb and the Engineering Ideal in Biology* (New York: Oxford University Press, 1987), pp. 65–69.

37. Child, *Individuality in Organisms.*

38. Child, *Origin and Development of the Nervous System,* p. 66.

39. Charles C. Adams, "Migration as a Factor in Evolution: Its Ecological Dynamics," *American Naturalist,* 1918, *52:* 465–490, see p. 472.

40. On the Chicago school of ecology, see J. Ronald Engel, *Sacred Sands: The Struggle for Community in the Indiana Dunes* (Middletown, Conn.: Wesleyan University Press, 1983).

41. Victor Shelford, "Physiological Animal Geography," *Journal of Morphology,* 1911, *22:* 1–38.

42. Garland E. Allen, *Thomas Hunt Morgan, The Man and His Science* (Princeton, N.J.: Princeton University Press, 1978).

43. Child, *Individuality in Organisms,* pp. 202–208.

44. Frederick B. Churchill, "Weismann, Hydromedusae, and the Biogenetic Imperative: A Reconsideration," in T. J. Horder, J. A. Witkowski, C. C. Wylie, eds., *A History of Embryology* (Cambridge: Cambridge University Press, 1986), pp. 7–33; idem, "Weismann's Continuity of the Germ-Plasm in Historical Perspective,"

Psychobiologist," *Transactions of the American Philosophical Society*, 1955, *45:* 1–85, see p. 68.

62. Herrick, "Clarence Luther Herrick," pp. 68–71; Jane Maienschein, "Whitman at Chicago: Establishing a Chicago style of biology?" in Rainger, Benson, and Maienschein, *The American Development of Biology*, pp. 151–184.

63. Herrick, "Clarence Luther Herrick," p. 7. Herrick attributed the coining of the term psychobiology to Adolf Meyer, psychiatrist at Johns Hopkins University, who was trying to integrate psychiatry and biology in the clinical setting. Meyer, however, never established a coherent theoretical approach to his science. See Adolf Meyer, "Objective Psychology or Psychobiology with Subordination of the Medically Useless Contrast of Mental and Physical," *Journal of the American Medical Association*, 1915, *65:* 860–862; Alfred Lief, *The Commonsense Psychiatry of Dr. Adolf Meyer* (New York: McGraw-Hill, 1948). On the relationship between C. J. Herrick's and Meyer's views, see Paul S. Roofe, "Neurology Comes of Age," *Journal of the Kansas Medical Society*, 1963, *66:* 124–129.

64. Bartelmez, "Charles Judson Herrick."

65. Herrick, "Clarence Luther Herrick," p. 11.

66. Ibid., p. 76, 77.

67. Adolf Meyer, "The Contemporary Setting of the Pioneer," *Journal of Comparative Neurology*, 1941, *74:* 1–24. See also C. J. Herrick, "The Founder and Early History of the Journal," *Journal of Comparative Neurology*, 1941, *74:* 25–38.

68. C. E. Coghill, "Clarence Luther Herrick as Teacher and Friend," *Journal of Comparative Neurology*, 1941, *74:* 39–42, see p. 41.

69. Robert J. Richards, *Darwin and the Emergence of Evolutionary Theories of Mind and Behavior* (Chicago: University of Chicago Press, 1987).

70. I have inferred this connection on the basis of Herrick's later comments about the similar experiences of his friend George E. Coghill: see C. J. Herrick, *George Ellett Coghill: Naturalist and Philosopher* (Chicago: University of Chicago Press, 1949), pp. 15–16. Compare Herrick's easy transition with the wholesale rejection of his Southern Baptist fundamentalist background by the founder of behaviorism, John Broadus Watson. See Paul G. Creelan, "Watsonian Behaviorism and the Calvinist Conscience," *Journal of the History of the Behavioral Sciences*, 1974, *10:* 96–118.

71. C. J. Herrick, "The Scientific Study of Man and the Humanities," in Leonard D. White, ed., *The New Social Science* (Chicago: University of Chicago Press, 1930), pp. 112–122.

72. C. J. Herrick, "A Biologist Looks at the Profit Motive," *Social Forces*, 1938, *16:* 320–327.

73. C. J. Herrick, "Clarence Luther Herrick," p. 76.

74. Adolf Meyer, "The Contemporary Setting of the Pioneer," *Journal of Comparative Neurology*, 1941, *74:* 1–24.

75. C. J. Herrick, *The Brain of the Tiger Salamander, Ambystoma tigrinum* (Chicago: University of Chicago Press, 1948).

76. Bartelmez, "Charles Judson Herrick," p. 87.

77. Ibid., p. 88. Members of the group included Percival Bailey, Paul C. Bucy, Stephen Polyak, David Bodian, A. Earl Walker, Karl S. Lashley, Heinrich Kluver,

from a symposium on "August Weismann (1834–1914) und die theoret des 19.Jahrhunderts," *Freiburger Universitätsblätter,* 1985, *87–88:* 1(

45. Child, *Individuality in Organisms,* pp. 204–205.

46. William B. Provine, *Sewall Wright and Evolutionary Biolo* University of Chicago Press, 1986), pp. 168–69.

47. Scott Gilbert, "Cellular Politics: Ernest Everett Just, Richa schmidt, and the Attempt to Reconcile Embryology and Genetics," in R. R. Benson, and J. Maienschein, eds., *The American Development of Bic* delphia: University of Pennsylvania Press, 1988), pp. 311–46. See als(*Beyond the Gene: Cytoplasmic Inheritance and the Struggle for Authority* (New York: Oxford University Press, 1987).

48. William B. Provine, "Francis B. Sumner and the Evolutionary : *Studies in History of Biology,* 1979, *3:* 211–240. See also the sympathetic cal memoir of Sumner by Child, "Francis Bertody Sumner, 1874–1945,' *Academy of Sciences Biographical Memoirs,* 1947, *25:* 147–173. On Wh Mary A. Evans and Howard E. Evans, *William Morton Wheeler, Biolog* bridge, Mass.: Harvard University Press, 1970).

49. Arthur M. Lewis, *Evolution, Social and Organic,* 5th ed. (Chicago H. Kerr, 1909).

50. In developing the analogy between the biological and social organisı drew on the work of Herbert Spencer and the sociological writings of the statesman and political economist Albert Eberhard Friedrich Schäffle (1831–

51. Child, *Individuality in Organisms,* p. 206.

52. C. M. Child, *Physiological Foundations of Behavior* (New York 1924), p. 268.

53. Ibid., pp. 287, 297–298.

54. Ibid., pp. 282, 289.

55. Child, *Physiological Foundations of Behavior,* p. 286.

56. C. M. Child, "The Individual and Environment from a Physiological ' point," in Jane Addams et al., *The Child, the Clinic and the Court* (New York: Republic, 1925; reprinted 1970), pp. 126–155.

57. Child influenced the sociological work of Roderick D. McKenzie and Dawson, for instance. On the influence of biological theories on sociology, reference to Child's impact, see Milla A. Alihan, *Social Ecology, a Critical Ana* (New York: Columbia University Press, 1938), pp. 108–118; Marlene Shore, *Science of Social Redemption: McGill, the Chicago School, and the Origins of So Research in Canada* (Toronto: University of Toronto Press, 1987), pp. 95–1 110–112.

58. William E. Ritter, *The Unity of the Organism,* 2 vols. (Boston: Gorha Press, 1919); Edward S. Russell, *The Interpretation of Development and Heredi* (Oxford: Clarendon Press, 1930).

59. Donna Haraway, "Animal Sociology and a Natural Economy of the Bod Politic, Part I: A Political Physiology of Dominance," *Signs,* 1978, *4:* 21–36.

60. Biographical information on Herrick is in George W. Bartelmez, "Charle: Judson Herrick, 1868–1960," *Biographical Memoirs of the National Academy o Sciences of U.S.A.,* 1973, *43:* 77–108.

61. C. J. Herrick, "Clarence Luther Herrick, Pioneer Naturalist, Teacher, and

Anton J. Carlson, Arno B. Luckhardt, Ralph S. Lillie, Ralph W. Gerard, Nathaniel Kleitman, Edmund Jacobson, Peter C. Kronfeld, Carl R. Moore, Paul Weiss, and B. H. Willier.

78. Richard W. Burkhardt, Jr., "The *Journal of Animal Behavior* and the Early History of Animal Behavior Studies in America," *Journal of Comparative Psychology,* 1987, *10:* 223–230.

79. Herbert S. Jennings, *Behavior of the Lower Organisms* (Bloomington, Ind.: Indiana University Press, 1962 [1906]).

80. H. S. Jennings, "Unsettled and Unsettling Problems in Science," Lecture No. 37, 11 March 1926, H. S. Jennings Papers, B:J44a, American Philosophical Society, Philadelphia. On the controversy between Jennings and Loeb, see Philip Pauly, *Controlling Life,* chap. 6.

81. Bartelmez, "Charles Judson Herrick," p. 79.

82. C. J. Herrick, *Neurological Foundations of Animal Behavior* (New York: Holt, 1924), p. 4.

83. Ibid., pp. 119–120.

84. Ibid., p. 294.

85. A good statement of the behaviorist position at this time is Karl S. Lashley, "The Behavioristic Interpretation of Consciousness," *Psychological Review,* 1923, *30:* 237–277; 329–353.

86. Herrick, *Neurological Foundations,* p. 306.

87. Mitman, *Evolution by Cooperation,* n. 7.

88. William M. Tuttle, Jr., *Race Riot: Chicago in the Red Summer of 1919* (New York: Atheneum, 1970).

89. Herrick, *Neurological Foundations,* p. 308.

90. On the influence of pragmatism in the social sciences and humanities at the University of Chicago while Herrick was there, see Egbert Darnell Rucker, *The Chicago Pragmatists* (Minneapolis: University of Minnesota Press, 1969).

91. Pauly, *Controlling Life,* pp. 41–45.

92. John Dewey, "The Need for a Recovery of Philosophy," in Dewey et al., *Creative Intelligence: Essays in the Pragmatic Attitude* (New York: Holt, 1917), p. 64. The literature on pragmatism is enormous. For an introduction to pragmatism, see Edward C. Moore, *American Pragmatism: Peirce, James, and Dewey* (New York: Columbia University Press, 1961); Israel Scheffler, *Four Pragmatists: A Critical Introduction to Peirce, James, Mead, and Dewey* (New York: Humanities Press, 1974). On Dewey's thought and career, see George Dykhuizen, *The Life and Mind of John Dewey* (Carbondale and Edwardsville: Southern Illinois University Press, 1973). An assessment of Dewey's liberalism made in the context of cold war conservatism is by George R. Geiger, *John Dewey in Perspective* (New York: Oxford University Press, 1958).

93. John Dewey, "Philosophy," in Charles Beard, ed., *Whither Mankind, a Panorama of Modern Civilization* (New York: Longmans, Green, 1928), pp. 313–331.

94. C. J. Herrick, "Behavior and Mechanism," *Social Forces,* 1928, *7:* 1–11.

95. Richards, *Darwin and the Emergence of Evolutionary Theories,* chap. 8 on C. Lloyd Morgan; William Morton Wheeler, *Emergent Evolution and the Development of Societies* (New York: Norton, 1928). Wheeler gave an address on emergent

evolution at a symposium at the Sixth International Congress of Philosophy, held at Cambridge, Mass., in 1926.

96. Herbert S. Jennings, "Diverse Doctrines of Evolution, Their Relation to the Practice of Science and of Life," *Science,* 1927, *65:* 19–25.

97. Ibid., p. 22.

98. Ibid., p. 25.

99. Richard W. Burkhardt, Jr., "Charles Otis Whitman, Wallace Craig, and the Biological Study of Behavior in the United States, 1898–1925," in Rainger, Benson, and Maienschein, *The American Development of Biology,* pp. 185–218; see the definition of biology quoted from Whitman, p. 188.

100. Francis B. Sumner, "The Naturalist as a Social Phenomenon," *American Naturalist,* 1940, *74:* 398–408.

101. Bertrand Russell, "Science," in Beard, *Whither Mankind,* pp. 63–82.

102. Silvan Schweber has argued for a distinctive American style in physics, reflecting a pragmatist tradition, in "The Empiricist Temper Regnant: Theoretical Physics in the United States, 1920–1950," *Historical Studies in the Physical Sciences,* 1986, *17:* 55–98.

103. Herrick, "The Scientific Study of Man and the Humanities," esp. p. 118.

104. Robert Kargon and Elizabeth Hodes, "Karl Compton, Isaiah Bowman, and the Politics of Science in the Great Depression," *Isis,* 1985, *76:* 301–318. Karl Compton is cited by C. J. Herrick in "A Biologist Looks at the Profit Motive." See also Ellis Hawley, "The Discovery and Study of a 'Corporate Liberalism,'" *Business History Review,* 1978, *52:* 309–320.

105. John Dewey, *Individualism, Old and New* (New York: Capricorn Books, 1962), p. 165.

106. Herrick, "A Biologist Looks at the Profit Motive," pp. 324–325.

107. Lawrence B. Davis, *Immigrants, Baptists, and the Protestant Mind in America* (Urbana: University of Illinois Press, 1973), pp. 1–8.

108. William Graebner, *The Engineering of Consent: Democracy and Authority in Twentieth-Century America* (Madison: University of Wisconsin Press, 1987), esp. chap. 1–2.

109. Pauly, *Controlling Life,* chap. 8.

110. C. J. Herrick, *The Evolution of Human Nature* (Austin: University of Texas Press, 1956), p. 189.

111. Richard Dawkins, *The Selfish Gene* (Oxford: Oxford University Press, 1976), chap. 11. The biological interpretation of fitness predominates in sociobiology, however.

112. Richards makes a similar point in contrasting modern sociobiology with Darwin's concept of human nature in *Darwin and the Emergence of Evolutionary Theories of Mind and Behavior.* Philip Kitcher has documented the ways that sociobiology has been misused in its applications to the human species. One can argue that these examples represent "popular science" or even "bad science," but they were published and presumably well funded. How do we draw the line between good science and bad science? Philip Kitcher, *Vaulting Ambition: Sociobiology and the Quest for Human Nature,* (Cambridge, Mass.: MIT Press, 1985).

Garland E. Allen

9

Old Wine in New Bottles: From Eugenics to Population Control in the Work of Raymond Pearl

Eugenics has been defined as the attempt to use theories of heredity to improve the genetic quality of the human species.[1] During the first three decades of the twentieth century, eugenics emerged as a widespread scientific and popular movement, both in Europe and the United States. Using the concepts of Mendelian heredity that had just been rediscovered, eugenicists sought to show that much of human social behavior was genetically determined. In practical terms, eugenicists wished to restrict the breeding of those individuals deemed to be socially inferior, and to encourage the breeding of those deemed to be socially superior. The eugenics movement had a widespread influence, particularly in the United States during the early decades of the twentieth century, where it provided a quasi-scientific rationale for passage of such legislation as the Immigration Restriction Act (Johnson Act) of 1924, and of eugenic sterilization and antimiscegentation laws by more than thirty states between 1907 and 1935.[2]

Various scholars writing on the eugenics movement in the United States and elsewhere have noted that its basic focus underwent a shift in the period after 1925 from an older, cruder hereditarianism to a newer, more sophisticated emphasis on the interaction between heredity and environment. Less well chronicled is the transformation of eugenic thought, in the same period, from concern with eugenics per se to concern with the larger issue of population control.[3] This concern emerged after World War II in the call for a curb on what appeared to be the unbridled expansion of the human population, especially in Third World countries and among the poor classes of society in the industrialized countries. As the literature of present-day population control organizations shows, every ill of modern society—from poverty to crime and pollution—is said to be the result of "too many people." Much of this general thinking had its origin in the old eugenics movement, especially in the United States, between 1910 and 1940. The shift was in the locus of social control by biological means. In the early decades of the century the locus was the individual family line—the Jukes, Kallikaks, or other "degenerate" germ plasms. By the late 1920s, however, the locus was the population: defined either nationally, racially, or in terms of social class. The general aim was the same: to use biological principles to understand and *control* social problems. It is in examining the shift in focus from the family lineage level (eugenics) to the large-scale population level (population control) that forms the basic thrust of this essay.[4]

To give coherence to the present discussion, I will focus on the work of one man, Raymond Pearl (1879–1940). For several reasons Pearl is a useful and important figure to illuminate the transition from old eugenics to new population control. First, Pearl was a well-known biologist with a considerable reputation both in the United States and abroad. Second, in the early decades of his career (1910–1925) Pearl was a strong and influential supporter of the eugenics movement. After the mid-1920s, however, he dissociated himself from the movement itself, severely criticizing its scientific basis. At the same time, however, he became one of the leading spokesmen in the United States on the issue of population growth and "overpopulation" of the world. Third, Pearl's switch from support of eugenics to support for population control was not random or capricious. His developing ideas show a clear ideological transition between eugenic and demographic thinking. In Pearl's mind the social value of each of these movements was the same: a desire to improve society through the use of known biological principles. In terms of social motivation and faith in modern science as a means of solving basic social problems, Pearl shows a direct continuity between the basic aims of eugenics and that of population control. As his support shifted from the former to the latter, Pearl's view of the causes of social problems did not change; what changed was the particular biological form in which he sought an explanation—and thus a solution.

Eugenics in early twentieth-century United States grew out of the convergence of several social and intellectual developments. The first is the whole collage of ideas and practices known generally as "progressivism." Although recent historians have questioned how cohesive progressivism was as a movement, all agree that certain broad changes in the fundamental area of social planning were taking place between 1890 and 1920. These included the idea of scientific management of social processes (commerce, the workplace, production levels, the economy) by middle-class, trained "experts," often claiming to be acting according to the laws of science.[5] Progressivism also included the general notion of planning for the future and of social control; that is, the management of human behavior toward more ordered productive ends. It was the opposite, in many ways, of laissez-faire and called specifically for an increased role by government and the state in the regulation of human affairs.[6] As the science of the "improvement of the human race through better breeding" (as C. B. Davenport put it), eugenics fit squarely into the progressive mold. It claimed a basis in legitimate science (Mendelian genetics) and developed a social program (immigration restriction, sterilization) that made the claim to eliminate the "dregs of society" in just a few generations. It was the application of a then new and exciting field of biology—genetics—to the solution of a wide range of persistent social problems, including alcoholism, pauperism, criminality, prostitution, rebelliousness, and feeblemindedness.

It is important to stress the compatibility between eugenics and progressivism, because many eugenicists thought of themselves as new scientific reformers.[7] Raymond Pearl and other biologists became involved out of a desire to apply their scientific expertise to a rational and long-term solution of persistent social problems. They saw themselves as differing from old-style social reformers and charity workers who blamed society and economic injustice for social problems, and wanted to change the social and economic system. To eugenicists and other "scientific" reformers, such solutions did not appear to attack the problem at its roots. The true cause of such problems lay not so much in defects in society as defects in individuals. To prevent those individuals from passing their defective traits to countless descendents constituted a meaningful and lasting solution. It is into this context—this self-image if you will—that Pearl and others sought to introduce biological arguments as a means of understanding and correcting some of Western society's most persistent social ills.

Early Life and Career of Raymond Pearl

Raymond Pearl was born at Farmington, New Hampshire, 3 June 1879, of an old New England family.[8] Educated at Dartmouth College (A.B., 1899) and the University of Michigan (Ph.D., 1902), Pearl showed an early interest in the use of mathematics and quantitative studies in biology. Later, while serving as an instructor in zoology at Michigan, Pearl took a year's leave of absence (1905–1906) during part of which he worked with biometrician Karl Pearson at University College (London) preparing a statistical study on assortative mating in protozoa. So readily did Pearl assimilate the biometric methods of Pearson, that in 1906 the latter invited him to become associate editor of the new journal *Biometrika*.

After returning to the United States in 1906, Pearl went first to the

University of Pennsylvania (1906–1907) and then to the Department of Biology at the Agricultural Experimental Station of the University of Maine, in Orono (1907–1918). During that time he carried out numerous studies on heredity and selection in various agriculturally important animals. Between 1917 and 1919 Pearl served as chief of the Statistical Division of the United States Food Administration. In 1918, while still employed in government services, he accepted a call from Professor William H. Welch to become professor of biometry and vital statistics in the newly formed School of Hygiene and Public Health at Johns Hopkins University. Pearl remained at Johns Hopkins in one capacity or another throughout the rest of his academic career. He was elected to the National Academy of Sciences in 1916 and served as a member of the academy's council from 1919 to 1925. From 1916 to 1918 he was a member of the executive council and chairman of the Agricultural Committee of the National Research Council. He was a member of the American Society of Zoologists (president in 1913), the American Philosophical Society, and the American Academy of Arts and Sciences.

Pearl's Early Work on Eugenics

Pearl's early association with Pearson and Francis Galton probably was responsible for his initial interest in both biometrics and biostatistics, on the one hand, and in eugenics, on the other. Galton and Pearson were leaders in both areas and in 1905–1906, while Pearl was in London, were in the process of organizing the Francis Galton Laboratory for the Study of National Eugenics. Pearson carried on Galton's work as head of the Galton Laboratory and as Galton Professor of Eugenics at University College well into the 1930s. The major thrust of this work was to unite eugenic concerns with biometrical methods of analysis.

Several of Pearl's early articles show clearly how the two subjects of eugenics and biometrics were closely intertwined in his thinking. In a biometrically oriented article of 1905, Pearl studied the correlation between brain weight and race.[9] And in a eugenically oriented article in 1908, he argued from biometric data that moral and mental traits were inherited and could be bred in or out of a population depending on the selective measures applied.[10] Thus, from the beginning Pearl approached eugenics as a biometrician rather than as a Mendelian, a factor which differentiated him from many of his colleagues in the United States. That is, as a biometrician Pearl worked with statistical distributions, correlations, and regressions, rather than with raw breeding data in the tradition of animal and plant hybridizers. Although he later accepted the Mendelian theory in full (unlike many of the biometricians in England), Pearl's initial approach to eugenics through biometrics was to influence his later move from eugenics into the theory of population control. Unlike his Mendelian, laboratory-oriented colleagues, Pearl came to eugenics with a populational perspective: that is, with the ability to think of selection acting on large-scale groups of organisms, rather than on individual breeding pairs.

In his 1908 article Pearl spelled out his early views on eugenics. He defined eugenics as "the science which deals with all influences that improve the inborn qualities of a race, also with those that develop them to the utmost advantage; and it embodies the study of agencies under social control that may improve or impair the racial qualities of future generations."[11] To Pearl, eugenic considerations were vital to the future of the human species. Traditionally, he pointed out, natural selection had operated to weed out the unfit, the so-called undesirable citizens of organic life, but human beings have been able to ameliorate the effects of the environment to such a degree that they have greatly reduced the action of selection. Pearl asserted that something must be brought into play to replace crude natural selection. And this, he claimed, "eugenics aims to do. The unfit, the great body of physical, mental and moral derelicts, must not be allowed to reproduce themselves indiscriminately as they now for the most part are. And further, every legitimate effort possible must be made to encourage the reproduction of the fittest."[12] Although Pearl admired social Darwinism, he could not accept the crudities and harshness of a completely laissez-faire attitude toward the "unfit." As he wrote in 1908: "Our highly developed human sympathy will no longer allow us to watch the state purify itself by aid of crude natural selection."[13] To Pearl, the value of eugenics was that it was based on science and scientific methods. "Hitherto," he wrote, "everybody except the scientist had a chance at directing the course of human evolution. In the eugenics movement an earnest attempt is being made to show that science is the only safe guide in respect to the most fundamental of social problems."[14] Eugenics was the rationalist approach that would direct human evolution along a positive course.

To Pearl, eugenics research must be able to answer two fundamental questions. (1) Are human *physical characters* inherited? (2) If physical characters *are* inherited, what about mental and moral characters? In his 1908 article Pearl used correlation coefficients from data collected by Galton and was able to answer the first question clearly and affirmatively. Many human physical characters are inherited in what appears to be a Mendelian pattern.[15]

The answers to the second question—are mental and moral traits inherited?—came from Pearson's work comparing "physiological characters" of brothers and sisters ("ability, temper, vivacity, assertiveness, conscientiousness, etc."). Alluding to the difficulties encountered in identifying and measuring such traits, Pearl noted: "It might appear at first thought that such characters could not be treated metrically, because no one of them can be measured in the individual with absolute accuracy. But such is far from the case. Developments of higher statistical theory make it possible to treat data of this kind with quantitative precision."[16] Pearl's statement is curious. It is obvious that the quantitative measurement of a given trait is quite different from the statistical analysis of measurements of that trait. No matter what sophisticated statistical techniques are available, nothing makes it possible to quantify accurately a subjectively defined trait. In his argument Pearl was trying to show that since physical traits, which can be accurately measured and which are known to be largely inherited, show high correlation coefficients between parent and child, the existence of similarly high correlation coefficients for mental and moral traits between parent and child must also be inherited. At best, he was reasoning by analogy.

Pearl went on to argue that the way to increase or decrease the presence of specific mental and moral traits in the human population must be through the same avenues as used for physical traits—selective breeding. Thus he favored the two-pronged approach popular among eugenicists in the early decades of the century: positive and negative eugenics. Positive eugenics involved encouraging the morally and mentally fittest individuals in society to have more children; negative eugenics involved discouraging the more morally and mentally "degenerate" individuals from having many or any children. Ultimately, Pearl hoped, the government could be persuaded to take this matter seriously and institute corrective procedures and programs on a large scale.[17] Pearl noted with enthusiasm that eugenicists were proposing to control and direct the evolutionary process itself.

Growing Biological Doubts about Eugenic Claims

Although he remained a vocal supporter of eugenics for another eighteen years, a number of biological considerations ultimately forced Pearl to reconsider his views on the effectiveness of the eugenic program. Several of these considerations were expressed by biologists in the period before World War I and had as much to do with questions about the effectiveness of natural selection and the relationship between genotype and phenotype, as with directly genetic issues of whether a given moral or personality trait was inherited. Pearl's own experiences as director of the agricultural station at the University of Maine brought him squarely into contact with issues about the effectiveness of selection and the distinction between phenotype and genotype, and thus to doubts about eugenics as a practical solution to social problems. What were Pearl's views on the role of selection and on the phenotype-genotype distinction, and how did these ultimately influence his views on eugenics?

Pearl's Doubts about the Effectiveness of Selection

Pearl, like many biologists in the first decades of the century, was as skeptical about the effectiveness of natural selection as a means of creating new species as he was convinced of the reality of evolution as a natural process. This skepticism distinguished him from other biometricians and may have preceded his work at the University of Maine Agricultural Station. However, between 1908 and 1910 Pearl's earlier skepticism was greatly reinforced by his own experience as director of a large-scale selection project at Orono. Pearl had inherited a long-term project at the station in which various methods were tried to increase egg-laying capacity in chickens. Between 1898 and 1908, when Pearl took over the project and did his first analysis of the data, there had been not only no increase in egg-laying capacity but actually a slight *decrease*.[18] Pearl did not believe that selection, natural or artificial, could produce significant and lasting change in a population. Most important, it could not produce a new type, or species. Furthermore, he was reinforced by the findings of Danish botanist Wilhelm Johannsen (1857–1927), whose selection experiments with beans around 1901–1903 had shown that selection ceases to produce significant change in a trait after about twelve to fifteen generations.[19] Many biologists, Pearl among them, concluded from Johannsen's work that selection was only a negative agent in producing real evolutionary change. Selection could weed out the unfit or separate an initially heterogeneous population into homogeneous, pure lines, but it could not produce anything new. Pearl opposed the view of W. E. Castle at Harvard, who claimed that selection actually increases variation in the direction in which selection is taking place.[20] Thus, according to Pearl, unless spontaneous variations for increased fecundity occur, selection is powerless to alter egg-laying capacity beyond a certain point.

The Genotype-Phenotype Distinction

In 1909 Johannsen offered the first clear distinction between the phenotype (appearance) and the genotype (actual genetic makeup) of an organism.[21] Johannsen pointed out that the phenotype of an organism was often deceptive in that it did not necessarily provide a clue as to what the organism could pass on to its offspring. Brown-coated mice all look the same, but some produce gray-coated and brown-coated offspring, whereas others produce only brown. One could not always determine the genotype from inspection of the phenotype. As Pearl interpreted Johannsen's important distinction, chickens who were good egg layers did not necessarily produce offspring that were good egg layers. Failure to understand the genotype-phenotype distinction could explain the poor results from ten years of selection work at the Maine Agricultural Station. The process of simply selecting the best egg layers from which to breed, referred to at the time as the "German method" or the method of "mass selection," was thus in error.

Pearl drew two conclusions from Johannsen's phenotype-genotype distinction: (1) Johannsen showed that Darwinian selection, in the wild or in the barnyard, was not an effective agent of long-term change because it selected only phenotypes when the important focus, at least for producing lasting change in a population, should be the genotype; (2) it suggested a new strategy for the egg-laying experiments, one in which he would breed not necessarily from the individual showing the desired phenotype but from one that could *transmit* the phenotype.[22] To Pearl, this view was entirely consistent with Mendelian principles as well as with Johannsen's pure-line experiments: both showed that the genetic composition (genotype) of an organism was not necessarily revealed by inspection of the outward appearance (phenotype).

By preserving individuals until he could determine the egg-laying capacity of their offspring, Pearl was able to reverse the downward trend in egg-laying capacity and actually produce an upswing in fecundity of the population as a whole.[23] However, by the end of the experiment (1915) egg-laying capacity of the population was not significantly higher than it had been at the very beginning of the project in 1898. The results confirmed Pearl's view that: (1) selection could not usually produce any significant change in overall makeup of the population; (2) when it did, it was by selecting for the *genotype,* not phenotype. Such a practice could only be carried out with hindsight—that is, after one had been able to inspect an individual's offspring; and (3) selection was at best only a negative factor, sorting out existing genotypes but not producing new ones.

The fact that Pearl could make a clear distinction between genotype and phenotype did not completely clarify his understanding of the principle of selection, especially when applied to natural populations. Noting that selection could not produce any lasting effects if it acted only on the phenotype, he put forward the curious argument that only selective agents that act *directly* on the germ cells could be of evolutionary importance. He went so far as to cite an experiment of his own in which he claims to have applied a selective agent (alcohol) directly to the germ cells of domestic fowl, from which he later bred.[24] The method of treating the germ cells was to have the parents inhale alcohol, breed from these birds, and determine the effects on several designated traits in the offspring, compared to a set of controls. (In an era when glue sniffing had become a popular fad, alcohol-sniffing chickens must have seemed something of an academic curiosity.)

Pearl's general conclusion was that "out of 12 different characters for which we have exact quantitative data, the offspring of treated parents (exposed to alcohol vapor) taken as a group are superior to the offspring of untreated parents in 8 characters."[25] These results may have been of popular interest in 1916, when agitation for prohibition was mounting, but their real significance for Pearl lay in how he interpreted the results. Alcohol, Pearl asserted, acted directly on the germ cells, "killing or inactivating the weak [germ cells] and leaving only the strong and resistant (germ cells) to produce zygotes and somata."[26] Pearl did not elaborate on his evidence for this claim. In fact, the general assumption that any drug administered through the respiratory or gastrointestinal system necessarily acts directly on the germ cells has never been supported by physiological data, either in Pearl's day or our own. Yet the factual correctness or incorrectness of Pearl's assumption is less important than the implications that he saw for understanding the role of selection. To Pearl, the only way in which selection per se could be a direct agent for producing evolutionary change was for it to act directly on the germ cells, not indirectly through the somatic cells.

If selection could not be considered a factor in evolutionary change, how *did* species-level adaptations and characters arise? In a later article Pearl listed four factors that, in the experience of practical breeders, appeared to be responsible for directional change and the production of new forms: (1) improved conditions of the environment (food, water, temperature, etc.); (2) hybridization, which, through heterosis (hybrid vigor) led to new varieties; (3) the purification of previously mixed races or varieties by selective sorting (à la Johannsen); and (4) mutations, leading in a single step to new varieties or even species.[27] The fourth factor was a particularly influential idea in Pearl's thinking and bears closer examination.

Pearl, like many of his contemporaries, failed to appreciate the role that small variations, such as the mutations found in Thomas Hunt Morgan's fruit flies, could play in bringing about Darwinian natural selection. Pearl, who was clearly abreast of the new work in Mendelian genetics, had little idea how the raw material on which natural selection could act was generated. As a result, he was greatly attracted to the basic ideas put forth in Hugo de Vries's mutation theory, published between 1901 and 1903 in a massive, two-volume work.[28] de Vries claimed to have shown that for a number of varieties of the evening primrose (*Oenothera lamarckiana*) offspring frequently demonstrated sudden, qualitatively different traits (leaf shape, flower arrangement, etc.) from their parents. These large-scale changes he called "mutations." In many cases the offspring were infertile when back-crossed to their parents but were fertile among themselves. Here were what appeared to be new species, formed in a single step. de Vries's theory received an enormous amount of attention between 1901 and 1915. For many biologists it served as an important alternative to the Darwinian theory of natural selection.[29] Not only did de Vries's theory appear to solve many problems traditionally associated with Darwinian theory (e.g., the time scale of evolution and swamping), it also suggested a concrete mechanism for the origin of variations. Mutations, whatever their cause at the germ-cell level, produced new and distinct varieties which would not be lost through intermixing and diluting as was thought to happen with the small, "individual variations" on which Darwin's theory was based.

In the period between 1900 and 1930, the efficacy of eugenics was based on the principle of selection—weeding out the unfit and preserving the fit. By 1917 or 1918 Pearl's antagonism to selection as a primary agent of biological change appears to have begun to affect his view of the efficacy of the eugenical program. Although the problem of selection was not the only or even primary issue that brought Pearl to criticize eugenics, it seems to have raised in his mind important scientific criticisms of eugenic methods that he had not considered before. Criticism of eugenic principles based on the selection theory opened the door to criticism on other scientific (mostly genetic) grounds as well.

Pearl's Major Critique of Eugenics and the Eugenics Movement, 1921–1927

In addition to a general loss of faith in the efficacy of selection, Pearl began to lose faith in the genetic basis of eugenics. By the postwar period, he had become aware that eugenics and its parent science, genetics, were drifting farther and farther apart. Eugenics was becoming less of a science and more of a propaganda effort. When it was proposed that the eugenicists hold their international congress in 1921 with the International Genetics Congress, Pearl wrote to William Bateson, "I feel very strongly that this would be a mistake. You know as well as I do that practically the eugenics workers and the genetics workers have not so much in common as logically they might be supposed to have. . . . I wish if you feel at all the way I do about it that you would write to Morgan expressing your views on the subject as strongly as you can."[30]

To Pearl, the fact that eugenics had increasingly lost touch with modern experimental and biometrical work was a betrayal of its basic philosophy. If eugenics was to serve any function, it was to provide a scientific basis for the rational control of human evolution. To the extent that eugenics got away from rationality (science) and became more and more a propagandistic enterprise, Pearl lost sympathy with it. While even in the early days (prior to World War I) eugenic thinking had never been perceived as objective or rational as its proponents had claimed, in the postwar period the claims became more exaggerated and the racism more overt. For example, eugenic claims advanced for immigration restriction between 1921 and 1924 were often blatantly xenophobic, villifying many ethnic and national groups as moronic, feebleminded, morally defective, and "troublemakers." Like his